Membrane-Bound Enzymes

ADVANCES IN EXPERIMENTAL MEDICINE AND BIOLOGY

Volume 1
THE RETICULOENDOTHELIAL SYSTEM AND ATHEROSCLEROSIS
Edited by N. R. Di Luzio and R. Paoletti • 1967

Volume 2
PHARMACOLOGY OF HORMONAL POLYPEPTIDES AND PROTEINS
Edited by N. Back, L. Martini, and R. Paoletti • 1968

Volume 3
GERM-FREE BIOLOGY–Experimental and Clinical Aspects
Edited by E. A. Mirand and N. Back • 1969

Volume 4
DRUGS AFFECTING LIPID METABOLISM
Edited by W. L. Holmes, L. A. Carlson, and R. Paoletti • 1969

Volume 5
LYMPHATIC TISSUE AND GERMINAL CENTERS IN IMMUNE RESPONSE
Edited by L. Fiore-Donati and M. G. Hanna, Jr. • 1969

Volume 6
RED CELL METABOLISM AND FUNCTION
Edited by George J. Brewer • 1970

Volume 7
SURFACE CHEMISTRY OF BIOLOGICAL SYSTEMS
Edited by Martin Blank • 1970

Volume 8
BRADYKININ AND RELATED KININS: Cardiovascular, Biochemical, and Neural Actions
Edited by F. Sicuteri, M. Rocha e Silva, and N. Back • 1970

Volume 9
SHOCK: Biochemical, Pharmacological, and Clinical Aspects
Edited by A. Bertelli and N. Back • 1970

Volume 10
THE HUMAN TESTIS
Edited by E. Rosemberg and C. A. Paulsen • 1970

Volume 11
MUSCLE METABOLISM DURING EXERCISE
Edited by B. Pernow and B. Saltin • 1971

Volume 12
MORPHOLOGICAL AND FUNCTIONAL ASPECTS OF IMMUNITY
Edited by K. Lindahl-Kiessling, G. Alm, and M. G. Hanna, Jr. • 1971

Volume 13
CHEMISTRY AND BRAIN DEVELOPMENT
Edited by R. Paoletti and A. N. Davison • 1971

Volume 14
MEMBRANE-BOUND ENZYMES
Edited by G. Porcellati and F. di Jeso • 1971

Membrane-Bound Enzymes

Proceedings of an
International Symposium
held in Pavia, Italy
May 29-30, 1970

Edited by
Giuseppe Porcellati
and
Fernando di Jeso
Istituto di Chimica Biologica dell'Universita di Pavia
Pavia, Italy

 SPRINGER SCIENCE+BUSINESS MEDIA, LLC 1971

DOI 10.1007/978-1-4614-4616-3

Originally published by Plenum Press, New York in 1971
MyCopy version of the original edition 1971

PREFACE

The present volume contains all the papers presented at the International Conference on Membrane-Bound Enzymes held at Pavia in May 1970. The publication of its scientific content has been made possible by the collaboration of many scientists who have taken part at the Symposium and who deeply and actively discussed the lectures which were delivered. In order to ensure rapid publication, however, the discussion will not be reported here.

The general subject of membrane-bound enzymic activity, its behavior, localization and regulation, was explored in depth from the standpoints of the various contributors in biophysics, biochemistry, cytology and pharmacology. Each session was briefly introduced by the session chairman's remarks about the field under discussion. At the end of the Conference, Dr.R.M.C.Dawson made some concluding remarks.

The meeting is considered to have been very successful. It certainly gave a further stimulus to biochemical and physiological research workers in this field of study. The editors express their thanks to the authors of the papers and to the Plenum Publishing Corporation for the prompt response which has enabled the rapid publication of the volume, and to the auditorium of the meeting, which was attended by more than one hundred research workers concerned with the problems of membrane biology.

We are happy to acknowledge the financial support of various organizations, which have been listed in another part of this book. The contribution of the Secretarial Staff of our Institute to the editorial work is also gratefully acknowledged.

Pavia, September 1970

Giuseppe Porcellati
Fernando di Jeso

ACKNOWLEDGEMENTS

Acknowledgement is gratefully made for the valuable financial support received from the following organizations :

Abbott, S.p.A.	Latina, Italy
Bayer Italia, S.p.A.	Milano, Italy
Bayropharm Italiana, S.p.A.	Milano, Italy
Biochemia, s.r.l.	Milano, Italy
Bracco Industria Chimica, S.p.A.	Milano, Italy
Farmaceutici Midy, S.p.A.	Milano, Italy
Hoechst Italia, S.p.A.	Milano, Italy
Industria Farmacologica Crinos, S.p.A.	Como, Italy
Italseber, S.p.A.	Milano, Italy
Laboratorio Biologico Zanoni, s.r.l.	Milano, Italy
Società Farmaceutica Lepetit, S.p.A.	Milano, Italy
Ormonoterapia Richter	Milano, Italy
Pfizer Italiana, S.p.A.	Roma, Italy
Laboratori Sigma-Tau, S.p.A.	Roma, Italy
Vister Terapeutici, S.p.A.	Como, Italy
Zambeletti, S.p.A.	Milano, Italy

Giuseppe Porcellati
Fernando di Jeso

CONTENTS

PARTICIPANTS AT THE MEETING

G.F.AZZONE - Istituto di Patologia Generale, Università di Padova, 35100 Padova, Italy.

O.BARNABEI - Istituto di Fisiologia Generale, Università di Ferrara, 44100 Ferrara, Italy.

K.R.BRUCKDORFER - Queen Elizabeth College, Department of Nutrition, University of London, London, England.

A.BRUNI - Istituto di Farmacologia, Università di Padova, 35100 Padova, Italy.

E.CARAFOLI - Istituto di Patologia Generale, Università di Modena, 41100 Modena, Italy.

G.CARIGNANI - Istituto di Chimica Biologica, Università di Padova, 35100 Padova, Italy.

P.CERLETTI - Istituto di Biochimica Generale, Università di Milano, 20133 Milano, Italy.

A.COLBEAU - Laboratoire de Biochimie Mèdicale, Faculté de Médecine et Pharmacie de Grenoble, Grenoble, France.

A.R.CONTESSA - Istituto di Farmacologia, Università di Padova, 35100 Padova, Italy.

R.M.C.DAWSON - Department of Biochemistry, Institute of Animal Physiology, Agricultural Research Council, Babraham, Cambridge England.

F. di JESO - Istituto di Chimica Biologica, Università di Pavia, 27100 Pavia, Italy.

P.GAZZOTTI - Istituto di Patologia Generale, Università di Modena, 41100 Modena, Italy.

F.GUERRIERI - Istituto di Chimica Biologica, Università di Bari, 70126 Bari, Italy.

B.HAMPRECHT - Max-Planck Institut für Zellchemie, München, Germany.

F.LYNEN - Max-Planck Institut für Zellchemie, München,Germany.

N.E.LOFRUMENTO - Istituto di Chimica Biologica, Università di Bari, 70126 Bari, Italy.

M.LORUSSO - Istituto di Chimica Biologica, Università di Bari, 70126 Bari, Italy.

S.MASSARI - Istituto di Patologia Generale, Università di Padova, 35100 Padova, Italy.

A.J.MEIJER - Laboratory of Biochemistry, University of Amsterdam, The Nederlands.

J.NACHBAUR - Laboratoire de Biochimie Médicale, Faculté de Médecine et Pharmacie de Grenoble, Grenoble, France.

C.NÜSSLER - Max-Planck Institut für Zellchemie, München, Germany.

P.PALATINI - Istituto di Farmacologia, Università di Padova, 35100 Padova, Italy.

S.PAPA - Istituto di Chimica Biologica, Università di Bari, 70126 Bari, Italy.

R.PETERS - Department of Biochemistry, University of Cambridge, England.

G.PORCELLATI - Istituto di Chimica Biologica, Università di Pavia, 27100 Pavia, Italy.

E.QUAGLIARIELLO - Istituto di Chimica Biologica, Università di Bari, 70126 Bari, Italy.

P.J.QUINN - Department of Biochemistry, Institute of Animal Physiology, Agricultural Research Council, Babraham, Cambridge, England.

B.ROELOFSEN - Biochemisch Laboratorium der Rijksuniversiteit, Utrecht, The Nederlands.

C.R.ROSSI - Istituto di Chimica Biologica, Università di Padova, 35100 Padova, Italy.

C.S.ROSSI - Istituto di Chimica Biologica, Università di Padova, 35100 Padova, Italy.

J.C.SKOU - Fysiologisk Institut, Aarhus Universitet, DK-8000 Aarhus C., Denmark.

G.L.SOTTOCASA - Istituto di Chimica Biologica, Università di Trieste, 34127 Trieste, Italy.

J.M.TAGER - Laboratory of Biochemistry, University of Amsterdam, The Nederlands.

R.TIOZZO - Istituto di Patologia Generale, Università di Modena, 41100 Modena, Italy.

V.TOMASI - Istituto di Fisiologia Generale, Università di Ferrara, 44100 Ferrara, Italy.

A.TREVISANI - Istituto di Fisiologia Generale, Università di Ferrara, 44100 Ferrara, Italy.

L.L.M. van DEENEN - Biochemisch Laboratorium der Rijksuniversiteit, Utrecht, The Nederlands.

P.M.VIGNAIS - Laboratoire de Biochimie Médicale, Faculté de Médecine et Pharmacie de Grenoble, Grenoble, France.

P.V.VIGNAIS - Laboratoire de Biochimie Médicale, Faculté de Médecine et Pharmacie de Grenoble, Grenoble, France.

G.ZANETTI - Istituto di Biochimica Generale, Università di Milano, 20133 Milano, Italy.

R.F.A.ZWAAL - Biochemisch Laboratorium der Rijksuniversiteit, Utrecht, The Nederlands.

THE INTERACTION OF SOLUBLE PROTEINS WITH LIPID INTERFACES

R.M.C. DAWSON and P.J. QUINN

Biochemistry Department, Agricultural Research Council

Institute of Animal Physiology, Babraham, Cambridge, U.K.

There is now a vast amount of evidence to show that virtually all the multi-enzyme systems which exist in cells contain phospholipids as an integral part of their structure and activity. Generally this evidence is as follows - the enzymic activity can be substantially reduced by extracting the complexes with organic solvents, e.g. aqueous acetone or by treating them with phospholipases (A or C) and largely restored by reacting them with aqueous suspensions of isolated phospholipids. Often there is little specificity about the structure of the phospholipid required although it has recently been claimed that phosphatidylserine is specific for the restoration of delipidated transport ATPase (Wheeler & Whittam, 1970). There is little evidence to show precisely why the phospholipids are essential components of such enzyme complexes. Our lack of knowledge is usually covered by all enveloping general statements, such as that they act as cement substances holding the individual enzymes with their active centres orientated towards one another so that the lipid provides a medium for electron flow within complexes and between complexes. However, these explanations would not necessarily apply to the requirement for lipids by individual particulate enzyme reactions e.g. lecithin as a co-factor for D-3-hydroxybutyrate-NAD oxidoreductase (Jurtshuk, Sekuzu & Green, 1963) or acidic phospholipids for protoheme ferrolyase (Sawada, Takeshita, Sugita & Yoneyama, 1969) and here it is possible that the phospholipid may produce some activating conformation change of the enzyme protein.

In general the various approaches which have been used to determine the ways in which proteins and lipids are bonded in naturally occurring lipoprotein complexes have given rather limited information. X-ray diffraction particularly when applied to the myelin lipoprotein membrane has given data on the organisation of the protein and lipid molecules (Finean & Burge, 1963). Optical rotatory dispersion measurements on lipoproteins are not easy to interpret although it would appear that membranes contain a substantial part of the peptide chain in the α-helical form and that these helices are probably located in a hydrophobic environment (Chapman & Wallach, 1968; Gordon, Wallach & Straus, 1969). Nuclear magnetic resonance measurements have suggested that in some membranes the proteins and lipids are interacting hydrophobically (Chapman, Kamat, de Gier & Penkett, 1968) although similar spectra from plasma lipoproteins indicate that there is not extensive hydrophobic association between lipid and protein (Steim, Edner & Bargoot, 1968) a conclusion which is, however, rather contrary to that deduced from other studies (Scanu, 1967).

Because of the limitations of the present physical methods available for examining natural lipoprotein structures it is necessary to resort again to the model, i.e. to examine the properties of greatly simplified systems such as the interaction of a purified protein and a single lipid and to use the knowledge gained as a help to understanding the more complex situations. Here one can use the lipid as a particle in a bulk phase e.g. bimolecular leaflets of phospholipids in a liquid crystalline form (smetic mesophase) or as a monolayer orientated at an air water interface. Although experiments with bulk phase lipids can yield useful information, in our experience, the monolayer has certain advantages as an experimental model. The molecules of lipid at the interface are orientated and arranged in a well understood way and their concentration can be readily altered by careful expansion or compression of the film. A number of surface parameters can be used to ascertain the extent and nature of the interaction with a soluble protein introduced into the bulk phase. Thus, an increase in the surface pressure of a film on adding protein is generally assumed to represent the penetration of part at least of space occupying 'volume' of the protein into the film, although this may not reach to the hydrophobic fatty acyl chains. Because of the various factors which contribute to the surface pressure exerted by a monolayer (the equation of state has kinetic, cohesive and electrostatic terms) it is clear that the 'volume' penetrating into a specified film may not always be proportional to the surface pressure increment. The 'volume' of protein penetrating into the monolayer will be related to the surface pressure increment by a factor which will depend on the force area curve of the lipid film and any specific interactions between the components. Measurement of the interfacial potential of a monolayer can also give evidence of protein-lipid interactions although the interpretation of any changes can be difficult. This is because this potential is the sum of the vertical components of the intrinsic dipoles of the lipid molecules, including the ionic dipoles of the head groups as well as the orientated dipoles of water molecules adjacent to the film. Any peturbation of the aligned lipid molecules or ionic interactions or displacement of water molecules by the adsorbing protein can affect these dipoles. Finally with the introduction of labelled proteins a direct measurement of the surface radioactivity of the adsorbed protein can be made (Quinn & Dawson, 1969a). This gives a direct measurement of the total amount of protein concentrated in the surface phase, although it gives no information as to its location.

The monolayer technique possesses the possible disadvantage that, unless the penetrating parts of the proteins are exactly the same size and shape as the film molecules, holes will tend to form in the unimolecular layer (Haydon & Taylor, 1963). To maintain the planar structure this means the hole will fill with air unless some compensating collapse or realignment of the fatty acyl chains can occur. Thus, unless this realignment occurs the adsorption energy may be changed somewhat by the energy required to form the hole. In the bimolecular leaflet on the other hand penetration is likely to lead to the planar structure being destroyed to avoid the hole filling with water. Thus, the lipid chains will collapse inwards owing to the strongly positive free energy change required for the formation of a hydrocarbon/water interface; beyond a certain point the bimolecular leaflet may no longer maintain its continuous structure and will break down into numerous micellar particles with curved surfaces.

DEPENDENCE OF ADSORPTION TO LIPID MONOLAYER ON PROTEIN CONCENTRATION

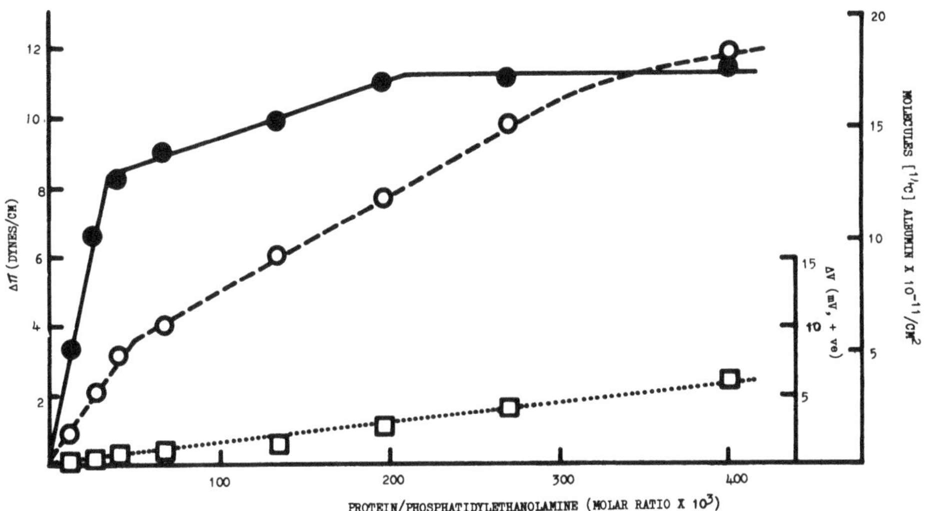

Fig. 1 Changes in surface pressure (•), radioactivity (o, ^{14}C-protein) and potential (□) on adding [^{14}C]carboxymethylated serum albumin below a monolayer of phosphatidylethanolamine at 2 dynes/cm.

When increasing concentrations of protein were added below a lipid monolayer which was initially at low pressure (2 dynes/cm) the surface pressure increments at equilibrium showed a relationship to protein concentration which was often biphasic (Fig. 1). The surface radioactivity change also showed a biphasic relationship but it is significant that whereas the pressure increment reached a maximum, that of the surface radioactivity continued to increase although presumably no additional penetration of the protein was occurring. It has often in the past been assumed that the protein added reacts quantitatively with the monolayer, but the surface radioactivity data showed that this was clearly not so. Quantification of the results indicated that at the minimum concentration of protein required to produce a maximum surface pressure increment then only a small percentage of the protein added had been adsorbed on the monolayer (Table 1). That the unadsorbed protein was not adsorbed on the surfaces of the trough in which the monolayer had been spread was indicated by the recovery of radioactivity in samples of the subphase taken from below the film. Furthermore, on sweeping off the monolayer of lipid containing the interacted protein (at equilibrium) between two barriers, surface active material accumulated at the interface as detected by radioactivity, pressure and potential or if a new lipid film was spread on such a freshly exposed subphase, the eventual increment in surface pressure and potential was approximately equal to the found with the initial film.

A comparison can be made of the true surface concentration of protein and the theoretical area made available by the compression of the lipid molecules by the increasing pressure caused by the penetration of the protein. This latter area can be deduced from the curve relating surface pressure and the area occupied by each lipid molecule in the absence of protein assuming that the pressure changes from the initial starting pressure (2 dynes/cm) to that at a point of inflection in the surface pressure/protein concentration curve. This 'area' made 'available' by the increase of pressure on the lipid molecule, can then be compared with the area occupied by the penetrating protein assuming that this is fully unfolded at the air water interface. This area of the protein can be calculated from its surface concentration (at the time the surface pressure

Table 1. Percentage of added protein bound to lipid monolayers

Protein	Lipid	First inflexion point	Second inflexion point
[^{14}C]Carboxymethylated cytochrome *c*	Stearic acid	74.0	26.0
	Phosphatidylcholine	10.0	2.1
	Phosphatidylethanolamine	26.0	18.8
[^{14}C]Carboxymethylated albumin	Stearic acid	20.5	15.4
	Phosphatidylcholine	19.4	11.4
	Phosphatidylethanolamine	14.0	6.6

Table 2. Stoichiometric relationship between proteins and lipids interacting at the air/water interface

A geometric method was used to compare the area occupied by the protein and the theoretical decrease in area of the lipid molecules at the interface.

		First inflexion point			Changes between first and second inflexion points		
Protein	Lipid	$\Delta\pi$ (dyn/cm)	$10^{14}\times$ Area of protein (Å^2/cm^2)	$10^{14}\times$ Decrease in area of lipid (Å^2/cm^2)	$\Delta\pi$ (dyn/cm)	$10^{14}\times$ Area of protein (Å^2/cm^2)	$10^{14}\times$ Decrease in area of lipid (Å^2/cm^2)
[^{14}C]Carboxymethylated cytochrome *c*	Stearic acid	6.4	7.8	6.0	3.6	4.4	2.0
	Phosphatidylcholine	10.8	25.4	19.4	5.2	21.3	4.0
	Phosphatidylethanolamine	10.8	23.1	22.0	2.0	29.4	3.4
[^{14}C]Carboxymethylated albumin	Stearic acid	4.1	5.1	4.3	3.0	9.3	0.8
	Phosphatidylcholine	8.1	18.6	16.5	2.9	28.9	2.9
	Phosphatidylethanolamine	8.5	16.4	17.8	2.6	17.8	5.1

becomes maximal) and a surface pressure/area per molecule curve obtained by spreading the same protein at an air/water interface. The calculated area occupied by the protein at the pressure appertaining at the first point of inflection showed a fairly close agreement to the area made 'available' by the compression of the lipid molecules considering the variety of experimental observations used in their derivation (Table 2). This indicates that, as originally suspected by Schulman (1957) the protein molecules in their entirety enter gaps in the expanded lipid films at low pressures and occupy spaces equivalent to that of the completely unfolded protein at the air water interface. Above this first inflection point the relationship broke down completely and it became clear that more protein was reacting with the film than could be accounted for by whole fully unfolded protein molecules entering the available space in the monolayer. This suggests that either the protein molecules were entering in a less unfolded form, or, alternatively, only part of their molecules, possibly a hydrophobic region, was entering the lipid film. It seems likely that at a low film pressure the energy required for protein molecules to enter the film and completely unfold at the interface must be comparatively low because of the greatly expanded nature of the monolayer, but that at higher pressures the adsorption energy required for this type of penetration is not available in the system. It should be noted that the first inflection point was below the collapse pressure of the protein at the air-water interface whereas the second was generally above this pressure. Whether whole protein molecules leave the lipid film after the first inflection point probably depends both on the nature of the protein and on the forces between it and the adjacent lipid molecules. With cytochrome c, for example, any conformational changes occurring would be expected to be reversible because of its well known resistance to denaturation. Thus, it will be seen later that, under certain circumstances, cytochrome c can be desorbed from lipid monolayers by compression, whereas with a protein which readily undergoes irreversible denaturation e.g. serum albumin, it can only be partially ejected by compression.

DEPENDENCE OF ADSORPTION TO LIPID MONOLAYER ON INITIAL PRESSURE OF FILM

In further experiments we have examined the change in surface parameters when the initial pressure of the lipid monolayer was varied and excess protein was added to the subphase.

Adsorption to Monolayers of Phosphatidylcholine

The simplest type of reaction is perhaps where the monolayer has head groups which are zwitterionic and uncharged at the pH of subphase used. On a subphase of 10mM sodium chloride penetration of cytochrome c into phosphatidylcholine monolayers as measured by an increase of surface pressure and the number of molecules adsorbing as judged by surface radioactivity were related linearly to the initial pressure of the monolayer having a negative slope and becoming zero at 20 dynes/cm (Fig. 2). None of the surface measurements gave any evidence of any interaction above this pressure. On adsorbing protein to the monolayer and then

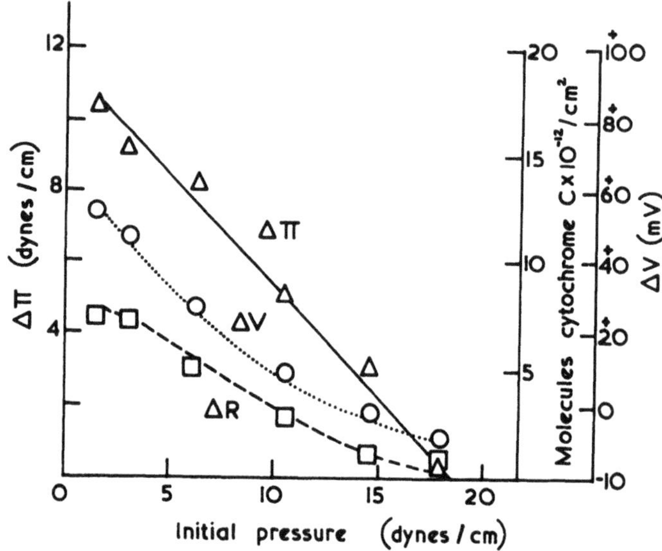

Fig. 2 Surface changes on adding cytochrome c below monolayers of egg lecithin at various initial pressures. $\Delta\Pi$=pressure(Δ), ΔV=potential (o), ΔR=radioactivity as ^{14}C-protein (□).

compressing to a pressure above 20 dynes/cm almost complete loss of surface radioactivity occurred representing total desorption. No difference was obtained if the adsorption was carried out in the presence of high concentrations of salt (NaCl 1M) confirming that ionic forces were having little effect on the interaction.

Adsorption to Monolayers of Acidic Phospholipids

The penetration of cytochrome c into monolayers of phosphatidylethanolamine showed a similar relationship to the initial pressure. However, in this instance, penetration (+ve surface pressure increment) occurred up to higher initial film pressures of 24 dynes/cm. It was apparent from the surface radioactivity determinations that in this system an interaction between protein and the lipid was occurring in the complete absence of penetration. At the subphase pH used (6), it is to be expected that the head group of the phosphatidylethanolamine molecule would be only partially negatively charged (due to limited depolarisation of the amino group). Using phosphatidic acid which is likely to possess at least one fully ionized anionic site on its head group at pH 6, this electrostatically facilitated adsorption without penetration was even more apparent (Fig. 3). Addition of M NaCl to reduce ionic bonding

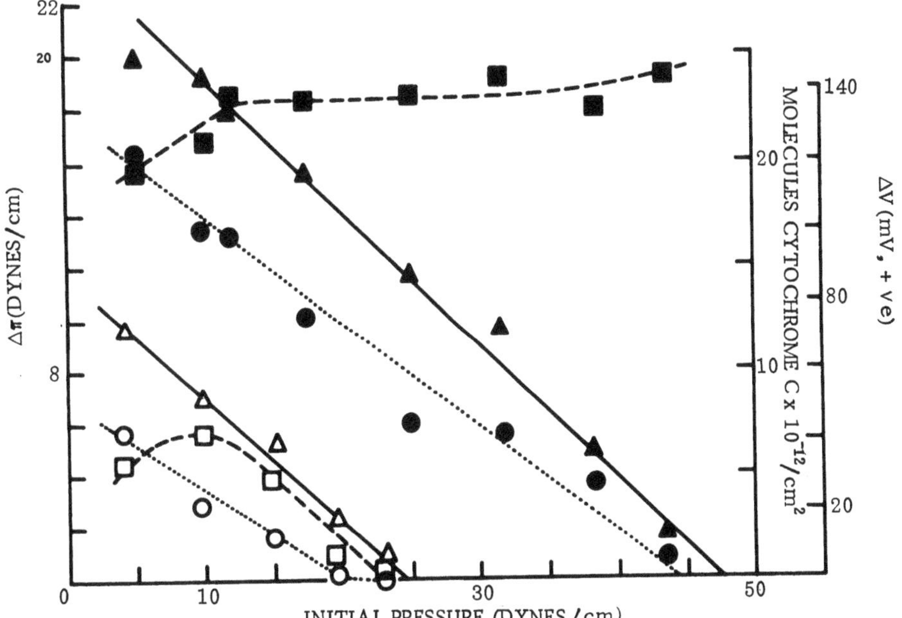

Fig. 3 Changes in surface pressure (▲), radioactivity (■, ^{14}C-protein) and potential (●), on adding [^{14}C]carboxymethylated cytochrome c below monolayers of phosphatidic acid at various initial starting pressures. Open symbols indicate that M NaCl has been added to the subphase.

suppressed the penetration so that a limiting pressure of only 24 dynes/cm was reached, similar to that into phosphatidylethanolamine monolayers. Furthermore there was now a complete absence of electrostatically facilitated adsorption without penetration. On compressing a film of phosphatidylethanolamine containing penetrated cytochrome c to 40 dynes/cm only a portion of the penetrated protein was ejected from the monolayer. However, if M NaCl was previously added to the subphase, practically all the penetrated protein was ejected, as was found with the uncharged phosphatidylcholine monolayers. This suggests that the penetration of protein into phospholipid films can be substantially facilitated and consolidated by electrostatic forces.

Penetration into Monolayers of Phosphatidylcholine Mixed with Anionic Lipids

The previous results had been obtained using a basic protein, cytochrome c which possessed a marked positive charge at the pH studied. To test whether this electrostatically-facilitated penetration was entirely dependent on the relative nett charges of the lipid-water interface and of the protein molecules in solution, measurements were made of the penetration of delipidated serum albumin into phosphatidylcholine monolayers alone or containing 10% molar stearylamine, dicetylphosphoric acid or phosphatidic acid. The subphase pH was varied between 4.5 and 5.5 which would produce a nett positive or nett negative charge on the albumin respectively, without causing any change of that on the lipid interface. The The results were unambiguous - penetration occurred into pure

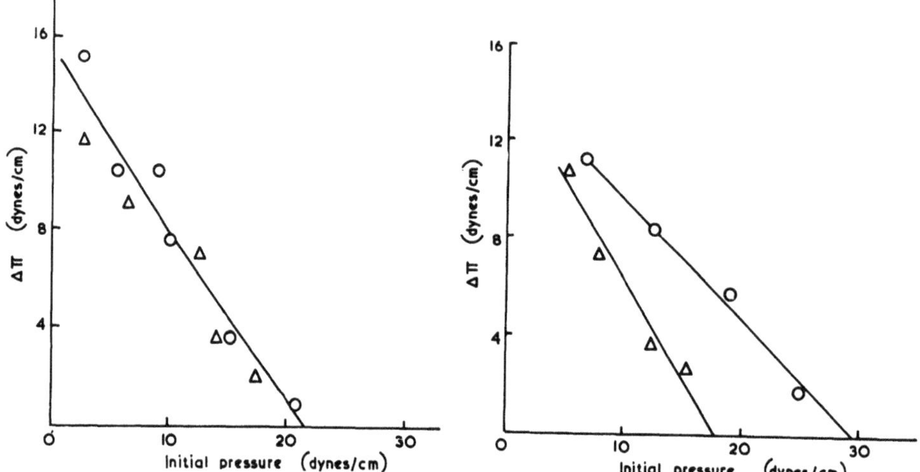

Fig. 4 Surface pressure increments (ΔΠ) on adding serum albumin below monolayers of yeast lecithin (left) and yeast lecithin + 10% molar phosphatidic acid (right) at various initial surface pressures. Subphase pH 4.5 (o), pH 5.5 (Δ)

phosphatidylcholine monolayers up to a pressure of about 20 dynes/cm (Fig. 4) irrespective of the subphase pH. Addition of the negatively-charged dicetylphosphoric acid or phosphatidic acid caused a marked increase in the penetration but only when the protein was positively charged (pH 4.5 subphase). In contrast, the inclusion of stearylamine in the film had little effect at either pH or even when the pH was increased to 7 to produce further additional anionic sites on the protein molecule. This would suggest that the electrostatically facilitated penetration may show some specificity for the sign of the nett charges on the protein and lipid interfaces. These results might provide an explanation for the specific activation of certain phospholipases by anionic amphipathic substances both in bulk and monolayer experiments (Dawson, 1968). The hydrolysis of monolayers of phosphatidylcholine by a deacylating enzyme of _P. notatum_ occurred only at low pressures, but at higher pressures a small proportion of long chain anions had to be added to the film (Bangham & Dawson, 1960). Since it seems likely that hydrolysis of the phospholipid would require extensive penetration of the active centre of the enzyme to reach the fatty acyl bonds, it is probable that this would be assisted by the anionic amphipath. This would be similar to the electrostatically facilitated penetration of serum albumin into phosphatidylcholine monolayers containing anionic amphipathic substances at pressures above 20 dynes/cm.

The explanation of this differential effect of the sign of the charge on the interface on penetration is obscure. It is perhaps not to be expected that all proteins would necessarily show the same charge assisted penetration characteristics, although experiments we have made with malate dehydrogenase suggest that this is similar to serum albumin. It has been observed that the binding of catalase (Fraser, Kaplan & Schulman, 1955) and trypsin (Fraser & Schulman, 1956) to oil emulsions was weak if the stabilizer was a positively charged detergent and strong if negatively charged detergents were used. It may also be relevant that the bulk of the phospholipids contained in biological membranes are zwitterions at a physiological pH while a small proportion possess a nett negative charge on their polar head groups (Dawson, 1968).

EFFECT OF FATTY ACID COMPOSITION OF PHOSPHOLIPID MONOLAYERS ON PENETRATION OF PROTEIN

The maximum pressure of monolayers of egg phosphatidylcholine and phosphatidylethanolamine allowing penetration of cytochrome c was not affected by hydrogenation of the lipids. The pressure increment/initial pressure curves showed slightly different slopes but the minimal initial pressure at which there was no penetration remained the same (Fig. 5). As expected the hydrogenation caused a pronounced reduction in the area occupied by each phospholipid at the interface e.g. at the penetration cut off pressure that of

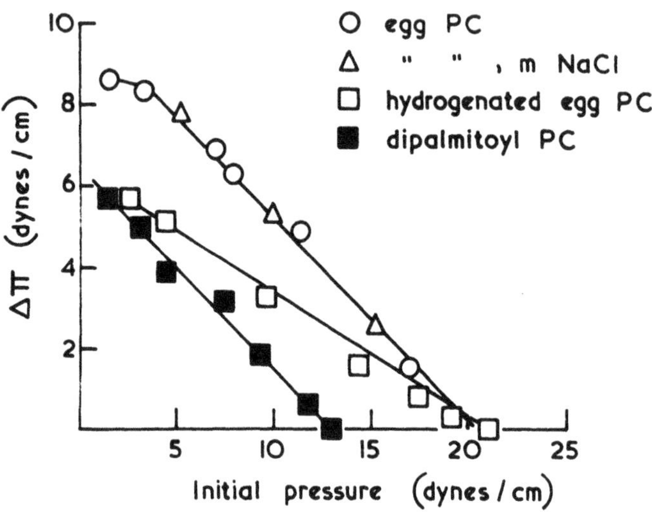

Fig. 5 Penetration of cytochrome c into phosphatidylcholine monolayers.

phosphatidylcholine had changed from 74A^{o2} to 50A^{o2} while that of phosphatidylethanolamine from 79A^{o2} to 47A^{o2}. Clearly the limiting pressure allowing penetration is not dependent on the space available between the lipid head groups. Presumably, for penetration to occur the forces assisting adsorption must be greater than the work required to push the occupying portion of the protein into the film so increasing the surface pressure. However, at lower pressures it is likely that for a specified change in the pressure increment on adding protein, a much smaller volume of the latter would enter the film of hydrogenated lipid due to the steeper slope of its force area curve.

Although unsaturation of the fatty acids appears to have little effect on the pressure allowing penetration, the latter is apparently enhanced by an increase in the length of the fatty acyl chains. Cytochrome c penetration into dipalmitoyl lecithin was cut off at 14 dynes/cm film pressure compared with 20 dynes/cm for hydrogenated egg lecithin (Fig. 5). The chain length here of the fatty acyl groups would be reduced by an average of 1 to 2 methylene units compared with the hydrogenated egg lecithin, although the area occupied by each molecule at the interface was practically identical. The effect was systematically examined with a series of lecithins of various fatty acyl chain length using delipidated serum albumin as the penetrating protein. Here the cut off point for penetration was about 20 dynes/cm for both dimyristoyl and dipalmitoyl lecithins but increased to 35 and 41 dynes/cm for distearoyl and dibehenoyl phosphatidylcholines respectively (Fig. 6). These lecithins show considerable differences in the properties of their monolayers, thus dibehenoyl and distearoyl lecithins are condensed

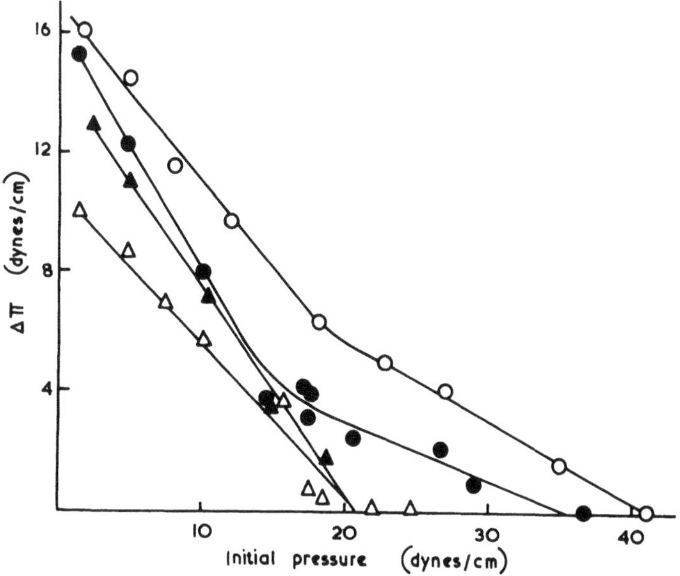

Fig. 6 Penetration (ΔΠ) of serum albumin into monolayers of saturated lecithins at various initial starting pressures Δ dimyristoyl, ▲ dipalmitoyl, ● distearoyl, o dibehenoyl

at room temperatures while dimyristoyl lecithin is liquid expanded. This again means that the pressure limit of protein penetration cannot be related to the gaps between the phospholipid molecules at the interface since dimyristoyl which has the most space available is not penetrated at pressures which allow penetration into distearoyl and dibehenoyl lecithins. Furthermore, there is no discontinuity in the protein penetration/initial pressure curve for dipalmitoyl lecithin where a transition occurs from the liquid expanded to the condensed state as the surface pressure is increased above 10 dynes/cm (Phillips & Chapman, 1968). Again it is to be expected that when penetration occurs, a greater volume of protein may have to enter the shorter chain lecithins to produce a given pressure increment.

What additional forces assist protein penetration into the longer chain lecithins at higher pressures is unknown. Direct hydrophobic interactions would involve the extensive penetration of unfolded areas of the protein since as pointed out by Haydon & Taylor (1963) the hydrophobic side chains of amino acids in the peptide chain would not be long enough to penetrate between the fatty acid chains of a phospholipid without some degree of unfolding. Again the pressure limit of penetration of albumin into yeast lecithin (whose fatty acids consist predominantly of palmitoleic acid and to a lesser extent oleic acid) was certainly no greater than that preventing penetration into saturated phosphatidylcholines of equivalent chain length, in spite of the former forming a highly expanded monolayer.

ADSORPTION OF PROTEINS WITHOUT PENETRATION

It has been indicated previously that with monolayers of acidic phospholipids, adsorption of albumin and cytochrome c could be detected by the surface radioactivity technique under conditions where there was no penetration. This adsorption is presumably electrostatically facilitated as it did not occur with phosphatidylcholine monolayers, or when M NaCl was added to the subphase. Although this non-penetrating adsorption undoubtedly occurs at all pressures it can be most conveniently studied with films at a pressure above the maximum permitting penetration. The adsorption of ^{14}C-cytochrome c to phosphatidylethanolamine monolayers at 25 dynes/cm was largely prevented by the inclusion of M NaCl in the subphase. However, if the adsorption was allowed to occur in 10mM salt and then solid NaCl (at arrow Fig. 7) was added to the subphase to bring the concentration to 1 M then only a portion of the adsorbed protein was displaced. The amount removed depended on the time of interaction with the film reducing to a constant level of about 20% at 40 mins and showing little variation with the initial pressure of the film. It is clear that once electrostatic interaction has occurred the adsorption is subsequently stabilized by nonionic bonding. This could be hydrogen bonding between the peptide chains and the polar head group of the phospholipid or less likely Van der Waals interactions between hydrophobic side chains of amino acids and the methylene groups of ethanolamine orientated coplanar to the surface.

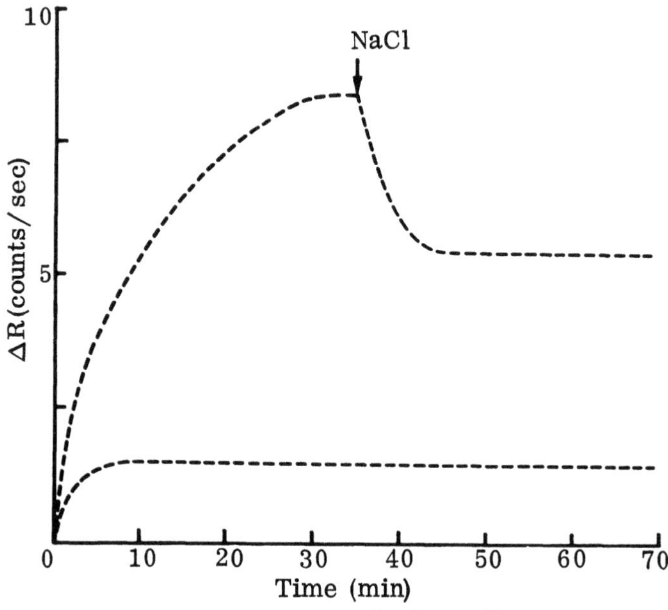

Fig. 7 Non-penetrating adsorption (ΔR) of ^{14}C-cytochrome c to phosphatidylethanolamine monolayers at 25 dynes/cm on 10mM NaCl (upper curve) or 1 M NaCl (lower curve)

EFFECT OF SUBPHASE pH ON PENETRATION AND ADSORPTION OF CYTOCHROME C

It can be predicted that the adsorption of proteins on phospholipid monolayers would be critically dependent on subphase pH, but the relationships we have found are complicated and imperfectly understood. Recently we have measured the penetration of simple peptides of known structure into monolayers, hoping to simplify the problems of interpretation involved but even here the results are far from clear cut.

It is intended here to discuss merely the results of the dependence of the binding of cytochrome c to phosphatidylethanolamine monolayers as a function of pH to illustrate the problems involved. Further details may be obtained from Quinn & Dawson (1969a,b). At high film pressures of phosphatidylethanolamine (26 dynes/cm) no penetration of cytochrome c occurred at any pH. The non-penetrating adsorption of the protein increased steadily as the pH was increased from pH 3 to about 9 (Fig. 8). This presumably corresponds to the relative electrostatic states of the protein and lipid interfaces

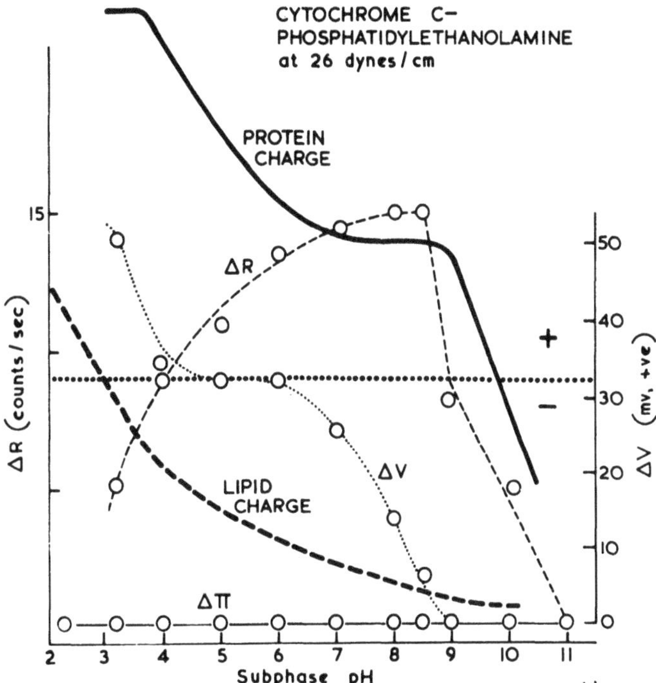

Fig. 8 Effect of subphase pH on the adsorption of ^{14}C-cytochrome c onto phosphatidylethanolamine films at 26 dynes/cm.
o —— o change in surface pressure, $\Delta\Pi$
o ··· o change in surface potential, ΔV
o --- o change in surface radioactivity, ΔR
To aid comparison, the charges of protein and phospholipid (Fig. 9) have been inserted with an isoelectric line but no scale.

and predominantly the latter. Micro-electrophoresis experiments suggest that the phospholipid-H_2O interface becomes progressively more negatively charged above pH 3 (Fig. 9). The protein's nett charge decreases largely due to the ionisation of the carboxylic acid groups but up to pH 9 it still possesses a significant nett +ve charge (Fig. 9). A precipitous decline in the adsorption of the protein to the monolayer occurred as the pH was increased above 9 and this coincides with the discharge of the ε amino groups of the 17 lysine residues in the protein (Fig. 8).

The effect of pH on penetration was examined by using a film of phosphatidylethanolamine at a lower pressure (10 dynes/cm) in the presence of M salt to avoid non-penetrating adsorption. This pressure is presumably high enough to avoid the penetration of whole completely unfolded protein molecules into the film. Here it is necessary to distinguish between the initial rate of reaction and the final equilibrium penetration. In spite of the presence of salt, there appeared to be some dependence of both on the relative electrostatic states of the protein and phospholipid (Fig. 10). Thus, the rate and final amount of penetration fell at pH 3 as the lipid interface became positively charged due to the depolarisation of the secondary phosphate groupings. The rates of penetration at pH

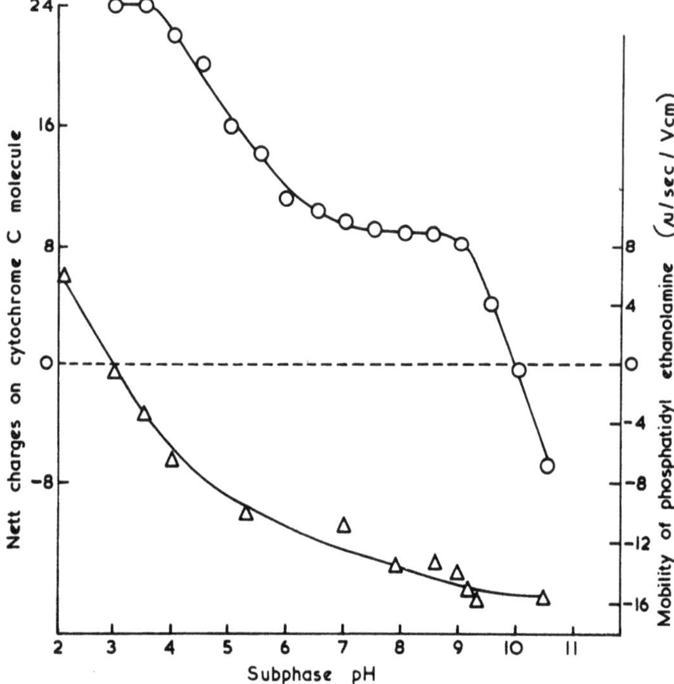

Fig. 9 Changes in the nett charge on cytochrome c molecules (o, by calculation) and on phosphatidylethanolamine particles (Δ, by microelectrophoresis) on varying the pH.

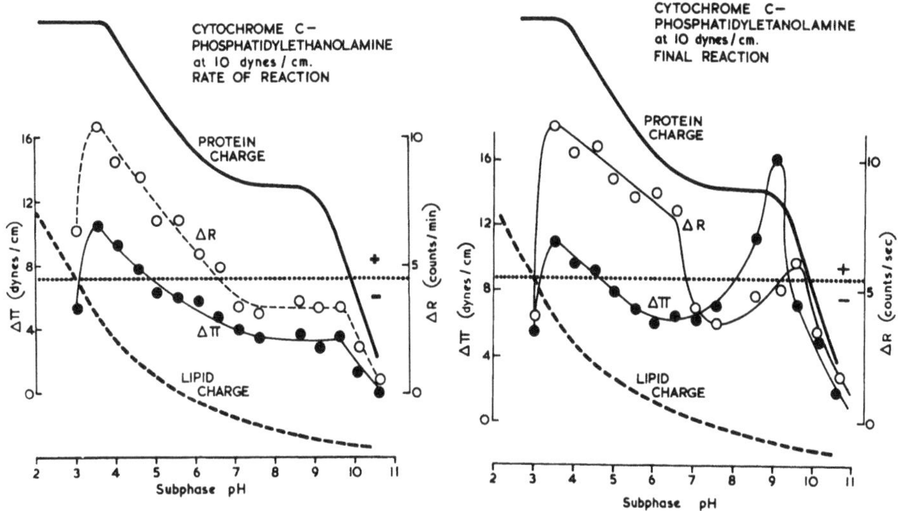

Fig. 10 Effect of subphase pH on the reaction between ^{14}C cytochrome c and phosphatidylethanolamine monolayers at 10 dynes/cm. M NaCl was present to avoid non-penetrating adsorption (Fig. 8)

o —— o change in surface pressure $\Delta\pi$ after 6 minutes (left) and on completion of the reaction (right)

o --- o change in surface radioactivity ΔR after 6 minutes (left) and on completion of the reaction (right)

To aid comparison, the charges of protein and phospholipid (Fig. 9) have been inserted with an isoelectric line but no scale.

values above 4 seem to correspond very closely to the nett positive charge on the protein (in contrast to non-penetrating adsorption) again showing a marked fall off as the ε amino groups of lysine were discharged. However, between pH 7 and 9 the final equilibrium penetration showed a marked difference to the initial rate, the pressure increment rising to a peak at pH 9. Moreover there was a clear departure from a constant ratio between the surface concentration of protein and the pressure increment generally seen at other pH values. This suggests that in this pH region conformational changes occur in the protein that allow more parts of the molecule to penetrate into the film. It is surprising that this corresponds to a pH region where there is little change in the ionisation of cytochrome c. As maximum penetration of cytochrome c into phosphatidylcholine monolayers also occurred at this pH, we originally suggested that it might be associated with conformational changes around the haem region which are known to occur between pH 7 and 9.5. However, subsequently we have observed the same high penetration just below the isoelectric point using ribonuclease where this explanation cannot apply.

The variation of the interaction of cytochrome c and a phospholipid surface with pH can clearly only partially be understood from the gross electrostatic factors involved and it is obvious that other subtle changes presumably in the conformation and hydration of the protein, considerably influence the magnitude and stability of the association. It is apparent that the often made assumption that the complex formation between cytochrome c and phospholipids can be described in purely electrostatic terms is an over simplification.

REVERSIBILITY OF PROTEIN ADSORPTION

Although the surface concentration of protein was clearly related to the concentration of protein added to the subphase (Fig. 1) the adsorption was essentially irreversible when the protein was removed from the bulk phase at maximum adsorption providing that the pressure was maintained. Thus, if protein was allowed to react with either a low pressure or high pressure monolayer between two barriers on the trough, and the film then slid to another compartment of the trough containing no protein in the subphase, there was no significant loss of film pressure or radioactivity (Table 3). With cytochrome c this applied whether the protein had penetrated or adsorbed without penetration. It appears, therefore, that protein reacted with films cannot rapidly desorb to come into equilibrium with a lowered concentration in the subphase and that the concentration of protein in some way determines the extent to which the energy barriers preventing penetration and bonding of the protein to the interface are overcome. The reaction apparently does not depend on the protein reacting with a fixed number of binding sites which are filled providing collisions are still occurring between protein molecules in the subphase and the lipid interface. Irreversible denaturation is unlikely to account for the irreversibility, because cytochrome c is remarkably resistant to such conformational changes and, moreover, the penetrated protein can be partially or wholly ejected by compression of the films to higher pressures. Further, it is clear that the protein remaining in the bulk phase does not differ in character to that bound, since on spreading a new film of lipid penetration of the remaining protein is almost as active as the original penetration.

The thermodynamic factors relating the adsorption to the protein concentration are not understood but, clearly if it is truly irreversible and does not desorb over long periods (Gonzalez & MacRitchie, 1970) the Gibbs equation relating adsorption on an interface to concentration may not apply since its derivation assumes reversibility of the adsorption.

Table 3. Effect of removing protein from the subphase at equilibrium on protein penetrated into or adsorbed on to phospholipid monolayers

Phospholipid monolayer	Protein	Surface measurement	Pressure (dyn/cm)	Radioactivity (c.p.s. – background)
Phosphatidylcholine	[14C]Carboxymethylated cytochrome *c*	Initial	2.1	0
		Final at equilibrium	12.2	35
		Transposed film after 45 min	12.5	31
Phosphatidylcholine	Delipidated albumin	Initial	2.0	—
		Final at equilibrium	17.1	—
		Transposed film after 45 min	17.0	—
Phosphatidyl-ethanolamine	[14C]Carboxymethylated cytochrome *c*	Initial	30.2	0
		Final at equilibrium	30.2	29
		Transposed film after 45 min	27.0	26

REFERENCES

Bangham, A.D. and Dawson, R.M.C., Biochem. J. 75:133 (1960).

Chapman, D., Kamat, V.B., de Gier, J. and Penkett, S.A., J. molec. Biol. 31:101(1968).

Chapman, D. and Wallach, D.F.H., in "Biological Membranes" (D. Chapman, ed.), p.125, Academic Press (1968)

Dawson, R.M.C., in "Biological Membranes" (D. Chapman, ed.), p.203, Academic Press (1968).

Finean, J.B. and Burge, R.E., J. molec. Biol. 7:672 (1963).

Fraser, M.J., Kaplan, J.G. and Schulman, J.H., Discuss. Faraday Soc. 20:44 (1955).

Fraser, M.J. and Schulman, J.H., J. colloid Sci. 11:451 (1956).

Gonzalez, G. and MacRitchie, F., J. colloid and Interfacial Sci. 32:55 (1970).

Gordon, A.S., Wallach, D.F.H. and Straus, J.H., Biochim. biophys. Acta 183:405 (1969).

Haydon, D.A. and Taylor, J., J. theoretical biol. 4:281 (1963).

Jurtshuk, P., Sekuzu, I. and Green, D.E., J. biol. chem. 238:3595 (1963).

Phillips, M.C. and Chapman, D., Biochim. biophys. Acta 163:301 (1968)

Quinn, P.J. and Dawson, R.M.C., Biochem. J. 113:791 (1969a)

Quinn, P.J. and Dawson, R.M.C., Biochem. J. 115:65 (1969b).

Sawada, H., Takeshita, M., Sugita, Y. and Yoneyama, Y., Biochim. biophys. Acta, 178:145 (1969).

Scanu, A., J. biol. Chem. 242:711 (1967).

Schulman, J.H., Discuss. Faraday Soc. 21:272 (1957).

Steim, J.M., Edner, O.J. and Bargoot, F.G., Science 162:909 (1968)

Wheeler, K.P. and Whittam, R., Nature 225:449 (1970).

THE MITOCHONDRIAL MEMBRANE AS A CHEMIOSMOTIC TRANSDUCER

G.F.Azzone and S. Massari

Istituto di Patologia Generale e Centro per lo

Studio della Fisiologia dei Mitocondri del CNR

About 30 years ago it was understood that the cell membrane had the capacity of translocating Na^+ and K^+. More recently this property has been related to the utilization of ATP (for a review cf Ref. 1). The conversion of the energy of a chemical compound such as ATP into an ion concentration gradient may be defined as a chemiosmotic transduction. It is immediately clear that this is a foundamental property of the plasma membrane. Very recently it has been reported (2) that the Na^+ pump of the red cell could be reversed, i.e. that ATP could be synthesized at the expense of concentration gradients of Na^+ and K^+. Unfortunately however the experiments gave only a qualitative indication of this reaction since no net synthesis of ATP could be detected.

The capacity of coupling the translocation of ion to the utilization of chemical energy is a basic property also of the inner membrane of animal mitochondria (3 -6). This property has been discovered much more recently but since then it has been investigated very intensively. For understanding the molecular mechanism of the chemiosmotic transduction, mitochondria represent a very convenient system to investigate. In fact they show a rapid response of their metabolic activity to changes in the rate of ion translocation (7).

The second important characteristic of the mitochondrial membrane, which is common to most natural membranes, is its capacity to form a barrier to ion diffusion (8). When a charged species, is observed to be translocated at a high rate across the mitochondrial membrane a specific mechanism of translocation is usually postulated (8).

We have now outlined two basic problems of mitochondrial function: a) the molecular mechanism for the chemiosmotic transduction and b) the molecular mechanism for ion translocation across an impermeable membrane. It is self evident that the two problems are closely interconnected and that the answer to the second problem is likely to be decisive for understanding the chemiosmotic coupling. We shall therefore discuss first this aspect especially in regard to the mechanism of translocation of univalent and divalent cations and of protons.

The information available is summarized in table I: a) some divalent cations such as Ca^{++}, Mn^{++}, Sr^{++}, can be rapidly translocated across the membrane while others such as Mg^{++} (except in muscle mitochondria), cannot; b) univalent cations cannot be translocated unless specific ionophores are added, such as valinomycin, monactin, nigericin; c) protons are quickly translocated when the membrane is energized.

Most critical is the question of the mechanism of proton translocation. Mitchell has been the first to recognize that the basic problem of the mitochondrial ion translocation process is that concerning the protons (8). In fact on one hand the lipid layer behave as proton impermeable (9) and on the other the energized mitochondria are able to extrude and take up proton at a very high rate (3). The contention of Mitchell has been disputed until very recently by Chance (10-11) who has maintained for a long time that the mitochondrial membrane is highly permeable to protons and therefore no special mechanism for proton translocation is needed. However this position appears no longer tenable particularly after the recent studies on artificial phospholipid bi-

TABLE I

Permeability to Ions of the Mitochondrial Membrane

	Ion	Permeability	Mechanism proposed
I^	Ca^{++}, Mn^{++}, Sr^{++},	High	Carrier
II^	K^{+}, Rb^{+}, Cs^{+}, Na^{+}, L_{i}^{+},	Low in absence and high in presence of ionophores	a) ionophores b) ionophores + carrier
III^	H^{+}	High under energized and low under deenergized conditions	a) none b) respiratory chain or ATPase c) uncouplers d) proton carrier $\pm$ uncouplers

layers which have shown a substantial impermeability of the lipid phase to protons and the H^+ conducting effect of the uncouplers (12-13). But then the question rises: how are the protons translocated through an impermeable membrane?

Mitchell has made a suggestion which has become widely known as the "chemiosmotic hypothesis" (8). This suggestion is illustrated in Fig. 1. According to Mit-

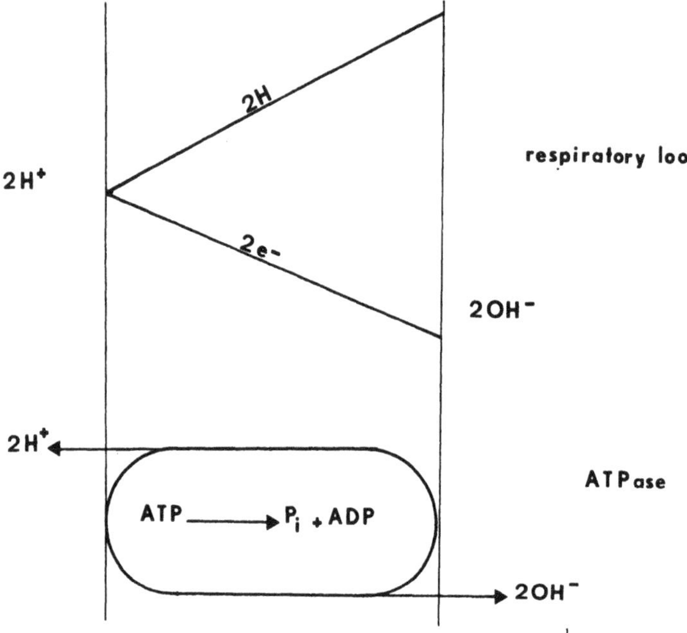

Fig. 1. Chemiosmotic hypothesis for H^+ translocation.

chell, the protons are translocated across the mitochondrial membrane in three ways: a) through the respiratory chain, b) through the ATPase and c) by means of proton conductors, i.e. the uncouplers. The respiratory chain of the mitochondria is seen by Mitchell as a sequence of hydrogen and electron carriers organized in three loops. As a consequence the oxidation of DPNH by oxygen results in the separation of 6 H^+ in the outer phase and 6 OH^- in the inner phase per oxygen atom consumed. The ATPase is also seen as a proton carrying system. The splitting of 1 equivalent of ATP is also accompanied by the translocation of two proton equivalents from the inner to the outer phase. According to this scheme therefore protons can be translocated through the membrane only when mitochondria are respiring or when they are splitting or synthesizing ATP. Alternatively the protons are let across the membrane when uncoupling agents are present.

At variance with this scheme we have proposed another mechanism (14-15) which is based on the presence in the membrane of a carrier catalyzing the electroneutral exchange of proton with cations, Fig. 2. This car-

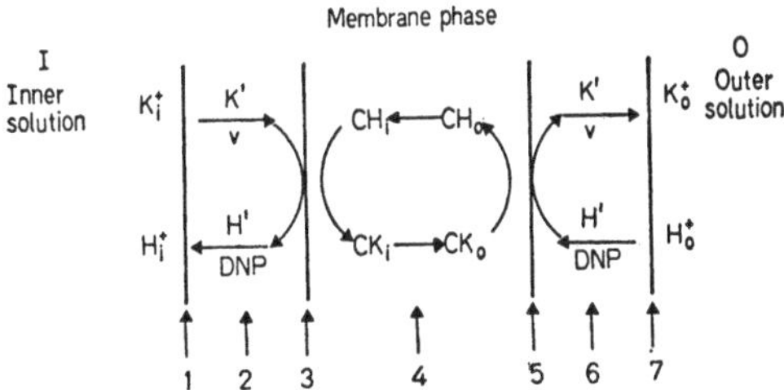

Fig. 2. Proton-cation carrier hypothesis for ion translocation.

rier has two fundamental properties: a) it has no direct accessibility to the univalent cation and protons present in the aqueous phase and is unable to translocate protons and cations in the absence of uncouplers and ionophores and b) when energized at the inner surface of the membrane the carrier has an increased affinity for protons, cf. Fig. 3; the energized carrier is able to catalyze a very

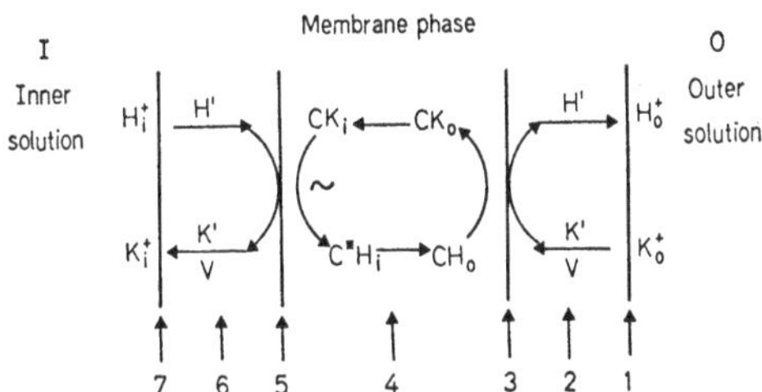

Fig. 3. Proton pump hypothesis for energy-linked H^+/K^+ exchange.

fast proton translocation.

The first property implies that the translocation catalyzed by the membrane carrier is in series with that catalyzed by the ionophores, such as valinomycin, and by uncouplers. In other words we postulate that in the mitochondrial membrane, at variance with the artificial membranes, valinomycin is unable to catalyze the transmembrane translocation of K^+ but simply lets the K^+ of the medium interact with the membrane carrier. Similarly uncoupling agents to not catalyze the transmembrane translocation of protons but let only the protons interact with the carrier.

The second property implies that when energy is available the carrier has an increased nucleophilicity at the inner side, i.e. the equilibrium constant for the exchange reaction of K^+ with H^+ is much larger. The anisotropy of this reaction provides the conditions for a source and a sink for vectorial reactions.

One possible objection against this mechanism is that an increase of nucleophilicity of the carrier due to metabolism is unable to provide a large supply of protons for the carrier translocation if the protons are to be taken only from the lipid phase which is assumed to have a low proton conductance. To this objection we can answer in two ways. First, although low, the conductance of the lipid phase for protons is not negligible and it may be sufficient for the activity of the energized carrier. Second, it is possible that the carrier activity is "gated". In other words the energization of the mitochondrial membrane brings about not only a change in the affinity constant for protons but also a variation of the accessibility of the carrier for the protons in the aqueous phase.

To distinguish between the two mechanism of proton translocation across the membrane we must consider two types of experiments. First, the kinetic analysis of the coupling of ion fluxes in the absence of metabolism. Second, the thermodynamic and kinetic analysis of the translocation of ions coupled to respiration, or to the synthesis of ATP.

ION TRANSLOCATION IN THE ABSENCE OF METABOLSIM

The translocation of ions across the mitochondrial membrane does not necessarily require metabolism. If we incubate mitochondria in a low K^+ medium, a gradient of osmotic energy will be available due to the high K_i^+/K_o^+ ratio. K^+ ions will flow down the concentration gradient and other ions, such as Ca^{++} or H^+ will flow in the opposite direction in a process which, taken microscopically, is electroneutral (16-17). Two explanations have been suggested in order to explain the coupling of the ion fluxes, cf Fig. 4. According to the chemiosmotic hypo-

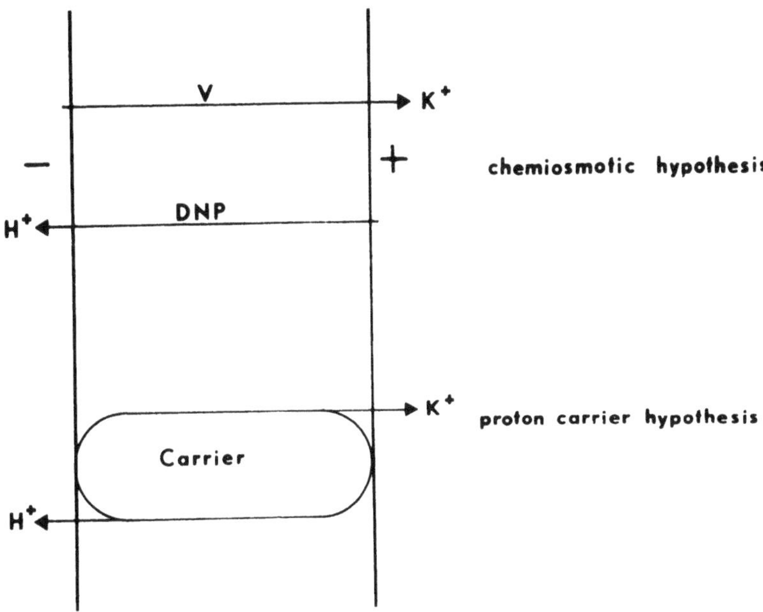

Fig. 4. Chemiosmotic and proton carrier hypotheses for coupling ion fluxes in the absence of metabolism.

thesis the efflux of K^+ via valinomycin is completely independent of the influx of H^+ via dinitrophenol, or of Ca^{++}. The only link between the two fluxes is the electrical potential created during the outward diffusion of K^+ ions. Therefore in this hypothesis although the macroscopic process is electroneutral the molecular process involves the separation of charges and the formation of a membrane potential. In the case of the proton carrier hypothesis, on the other hand, the efflux of K^+ and the influx of H^+ or of Ca^{++} are linked through the operation of a common carrier. No membrane potential is thus formed and the process is electroneutral also at the molecular level. We have elaborated a kinetic model which accounts for the rates of the ion translocation at the various concentrations of K^+, H^+, valinomycin and dinitrophenol. There is qualitative agreement between the theoretical equations and the observed rates. Furthermore a typical saturation kinetics has been obtained, as predicted by our model, in the plot of the rate of K^+ translocation vs the concentration of valinomycin, cf. Fig. 5.

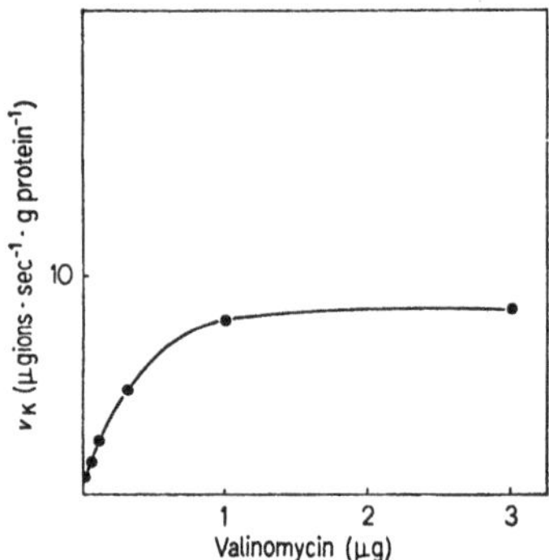

Fig. 5. Rates of K^+ release in the presence of uncouplers at various valinomycin concentrations. From Massari and Azzone (14), Fig. 11.

This saturation curve is difficult to explain on the basis of an electrogenic diffusion of K^+ across the membrane as predicted by the chemiosmotic hypothesis and strongly suggests a fixed number of sites to which the ions become bound during the translocation. It should be added in this regard that such a saturation kinetics is observed only when the membrane is not damaged, since otherwise the rate of K^+ uptake will increase more or less linearly with the increase of the concentration of the ionophore, Fig. 6. However, a limited penetration of the

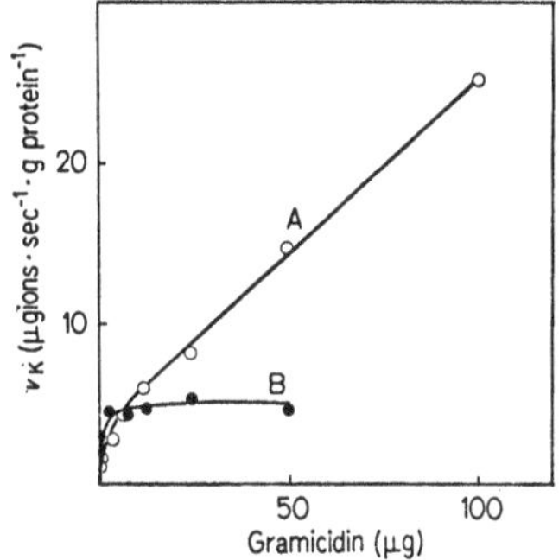

Fig. 6. Rates of K^+ release at various gramicidin concentrations. From Massari and Azzone, (14), Fig. 12. The curves A,B, were obtained by adding gramicidin in two ways: curve A, gramicidin was dissolved in an amount of ethanol variable from 0.1 to 100 μl (constant concentration 1 mg/ml); in curve B, gramicidin was dissolved in a constant volume of 10 μl ethanol (concentration variable from 0.02 to 5 mg/ml) 0.5 mM KCl.

K^+ valinomycin complex through the lipid phase, due to lack of neutralizing negative charges, cannot be excluded.

ION TRANSLOCATION COUPLED TO METABOLISM

The active uptake of cations is seen in a very different way in the chemiosmotic and proton pump hypotheses. In the former (8) the electron transport through the respiratory chain or the ATPase activity set up a membrane potential, while the permeable cation flow down the electrical gradient to reach electrochemical equilibrium. In the latter (15), the carrier binds protons at the inner side of the membrane and then transfers the protons at the outer surface where it exchanges protons with K^+. In this mechanism the cations are not at thermodynamic equilibrium, and no membrane potential is set up across the membrane. Also for the metabolism linked translocation we have set up kinetic equations and tested their compatibility with the observed rates of K^+ translocation. Phenomena of saturation and of competition have been observed in agreement with the prediction of the model.

One experiment, particularly relevant for discriminating between the two hypotheses, concerns the stoicheometry of the ion translocation with the oxygen uptake. According to the chemiosmotic mechanism the respiratory linked extrusion of H^+ occurs because of the operation of hydrogen carriers and there must be as many carriers (or half as much) as there are protons extruded per oxygen atom consumed. In agreement with the prediction of the model, Mitchell and Moyle have shown (18) that the H^+/O ratio for the NAD linked substrates is 6.

However, as we have pointed out, it is not correct to calculate the stoicheometry of the H^+ translocation on the basis of pH changes because the respiration linked H^+/K^+ exchange is accompanied by an anion uptake and this requires an OH^-/A^- exchange. We have therefore suggested to calculate the H^+/O ratio taking the cation/oxygen ratio, in view of the observation that the overall process of cation uptake is electroneutral (19). A detailed investigation of this process (20-21) carried out in our laboratory (Fig. 7) has revealed that: a) the K^+/O ratio is not constant but depends on the external and internal K^+ and H^+ concentrations and also on the amount of valinomycin added and b) that the K^+/O ratio can reach values as high as 12 for NAD linked substrates. Both phenomena are not compatible with the actual formulation of the chemiosmotic hypothesis. First, the chemiosmotic hypothesis predicts a constant stoicheometry, independent of the concentrations of H^+, K^+, and of valinomycin. Second, the chemiosmotic hypothesis can let 6 H^+ go across the membrane per oxygen atom but not 12.

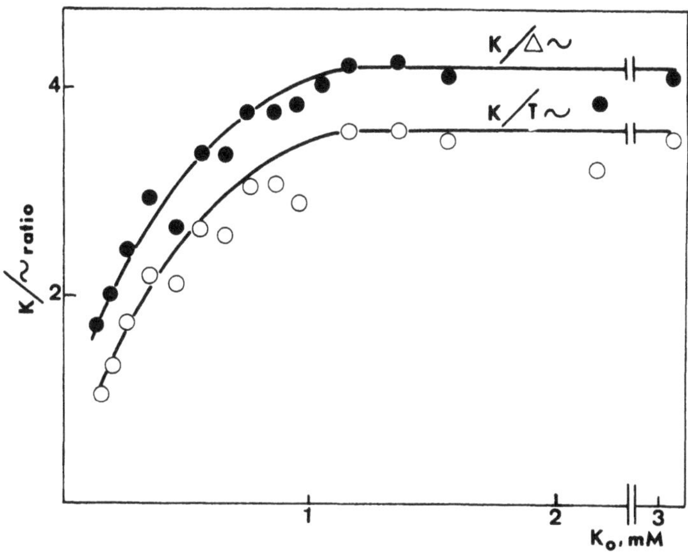

Fig. 7. K/~ ratio at various K_o concentrations. From Azzone and Massari, (21).

ION TRANSLOCATION COUPLED TO THE SYNTHESIS OF ATP

A large K^+ concentration gradient is present in mitochondria incubated in a low K^+ medium. Since the internal pH of mitochondria is usually 1.2 - 1.5 pH units higher than the external pH, a large proton concentration gradient is also present. The efflux of K^+ and the uptake of H^+ down their concentration gradients can be coupled to the synthesis of ATP (22-23). The experiment is carried out as follows (23). Mitochondria are incubated aerobically in the presence of substrates and P_i, Fig. 8. Addition of valinomycin initiates the K^+ uptake, which is accompanied by a stimulation of the respiration. After reaching a steady state, respiration is blocked by the addition of rotenone. This results only in a slow release of K^+. After a few seconds are added hexokinase and glucose, and ADP. ADP accelerates the K^+ efflux. During the phase of ADP stimulated K^+ efflux there is a large synthesis of ATP. We have studied in great detail

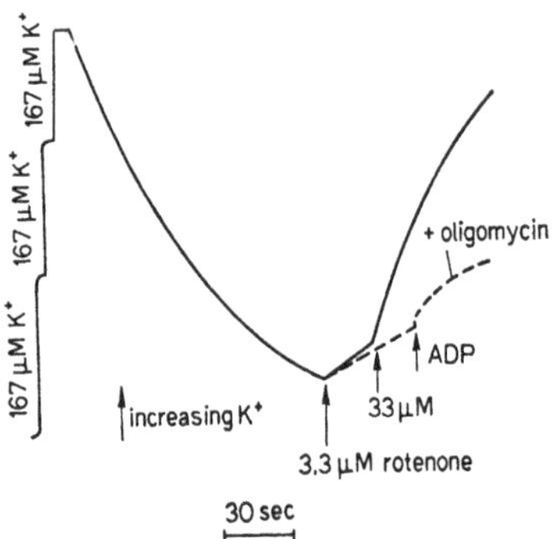

Fig. 8. Stimulation by ADP of K^+ efflux. From Rossi and Azzone (23). Fig. 1. K^+ uptake was initiated by the addition of 0.5 μg valinomycin. K^+ electrode traces are shown; the lower trace (dotted line) represents an experiment where 20 μg oligomycin was added 20 sec before rotenone.

this reaction from the point of view of its dependence on the concentrations of K^+, H^+, anions, ADP, hexokinase and also in respect to the stoicheometric relationship between the K^+ release and the synthesis of ATP (21 -23).

According to the chemiosmotic hypothesis the synthesis of ATP is due to the membrane potential formed during the outward diffusion of K^+. This potential drives the electrogenic ATPase as depicted in Fig. 1. According to our hypothesis the synthesis of ATP is essentially due to a reversal of the proton pump Fig. 9. The carrier becomes energized when translocates H^+ from the outer to the inner surface and then transfers the energy to an high energy intermediate involved in the synthesis of ATP. In the absence of ADP the reaction cannot proceed at the carrier is unable to continue the H^+/K^+ exchange. There are two experiments which are relevant for understanding the molecular mechanism of the reaction. One regards the requirement of H^+ influx during the K^+ efflux. If another permeant cation is added to the medium, for example Na^+ in the presence of monactin, the synthesis of ATP is strongly inhibited. This because Na^+ replaces

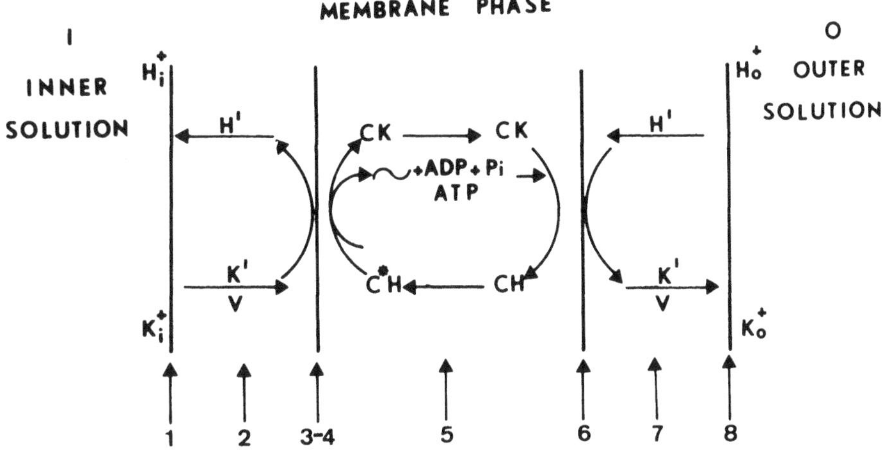

Fig. 9. Proton pump hypothesis for the synthesis of ATP coupled to H^+/K^+ exchange. From Azzone and Massari (21).

H^+, as can be measured also with the H^+ electrode. This experiment cannot be used to discriminate between the chemiosmotic and the proton pump hypothesis since both rely on a H^+ influx for the synthesis of ATP.

The other experiment regards the dependence of the ATP synthesis on the K_i^+/K_o^+ ratios. Fig. 10 shows that the amount of ATP synthesized decreased with the decrease of the K_i^+/K_o^+ ratio. The intercept with the abscissa gives the minimal K_i^+/K_o^+ ratio compatible with ATP synthesis. It is seen that this intercept was smaller at the more acidic pH. Since the Δ pH is larger at the more acidic pH this experiments indicates that the energy required to drive the ATP synthesis is provided by both the H^+ and K^+ concentrations gradients. By carrying out these experiments at various K^+ and H^+ concentrations gradients it has been found that the minimal amount of osmotic energy, as given by the sum of Δ pH + Δ pK, compatible with the synthesis of ATP remains roughly constant at the level of 2.5 units. Since the number of H^+/K^+ exchanges per ATP synthesized has been found to be

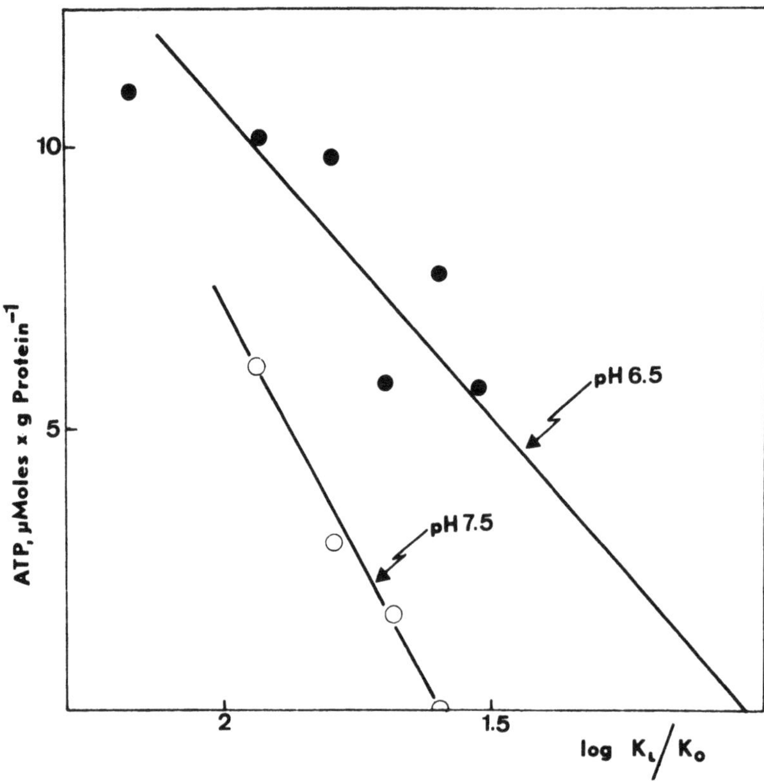

Fig. 10. Effect of K_i/K_o ratio on the synthesis of ATP at various pHo. From Azzone and Massari (21).

4, the minimal differences of osmotic energy compatible with the synthesis of ATP is 13.8 kcalories.

We have therefore here another example of disagreement between chemiosmotic hypothesis and experimental data. On one side the chemiosmotic hypothesis does not predict the translocation of 4 charges per ATP synthesized. On the other the translocation of only two charge as predicted by the chemiosmotic hypothesis would provide a difference of osmotic energy only of 6.9 kcalories which is about half that required for synthesizing ATP.

CONCLUSIONS

In this presentation we have summarized some experimental data which permit to discriminate between the two major hypotheses for the chemiosmotic transformations in mitochondrial membranes. It appears that the proton pump hypothesis explains a larger number of experimental findings than the chemiosmotic hypothesis. This is obviously not a proof for the validity of the proton pump hypothesis. An hypothesis created to explain experimental data can never be "proved" but only found to be "compatible". It is possible that the higher degree of "compatibility" of the proton pump hypothesis is due to its more recent presentation and therefore its not having passed yet the crithical test of the investigators.

REFERENCES

1. Caldwell P.C., Physiol. Rev. 48: 1 (1968).
2. Garrahan P.J. and Glynn I.M., Physiol. 192: 237 (1968).
3. Saris L., Soc. Sci. Fennica, Commentationes Biol. 28: 11 (1963).
4. Vasington F.D. and Murply J.V., J. Biol. Chem. 237: 2670 (1967).
5. Chappell J.B., Greville G.D. and Cohon M., In Energy Linked Functions of Mitochondria, p. 219, B.Chance ed. Academic Press, New York (1963).
6. Moore C. and Pressman B., Biochem. Biophys. Res. Commun. 15: 562 (1964).
7. Chance B., J. Biol. Chem. 240:2729 (1964).
8. Mitchell P., Biol. Rev. 41: 445 (1966).
9. Mitchell P. and Moyle J., Bhiochem. J. 104: 588 (1967).
10. Chance B. and Mela L., Nature 212: 375 (1966).
11. Lowestein J.M. and Chance B., J. Biol. Chem. 243: 3940 (1968).
12. Skulacev V.P., Sharof A.A. and Liberman E.A. Nature, 216: 718 (1967).
13. Bielawski J., Thomson T.E. and Lehninger A.L., Biochem. Biophys. Res. Commun. 24: 948 (1966).
14. Massari S. and Azzone G.F., Europ. J. Biochem. 12: 301 (1970).

15. Massari S. and Azzone G.F., Europ. J. Biochem., 12: 309 (1970).
16. Rossi C., Azzi A. and Azzone G.F., J. Biol. Chem., 242: 951 (1967).
17. Scarpa A. and Azzone G.F., Europ. J. Biochem., 12: 328 (1970).
18. Mitchell P. and Moyle J., Nature 208: 247 (1965).
19. Rossi C., Scarpa A. and Azzone G.F., Biochemistry, 6: 3902 (1967).
20. Rossi E. and Azzone G.F., Europ. J. Biochem., 7: 418 (1969).
21. Azzone G.F. and Massari S., to be published.
22. Cockrell R.S., Harris E.J. and Pressman B.C., Nature, 215: 1487 (1967).
23. Rossi E. and Azzone G.F., Europ. J.Biochem., 12: 318 (1970).

ANION TRANSLOCATION SYSTEMS OF THE INNER MITOCHONDRIAL MEMBRANE

E. Quagliariello, S. Papa, N.E. Lofrumento, A.J. Meijer and J.M. Tager

Department of Biochemistry, University of Bari (Bari) Italy and Laboratory of Biochemistry, University of Amsterdam (Amsterdam) The Netherlands

The inner mitochondrial membrane is not freely permeable to various anionic substrates. From time to time it has been proposed that the translocation of anionic substrates across the inner mitochondrial membrane is mediated by specific systems (1-3) through exchange-diffusion reactions of the respective anions with hydroxyl (2,4,5) or, what amounts to the same, through proton-anion symports (6,7). The occurrence of anion-anion exchange-diffusion reactions has also been postulated (2,3).

In this paper a study of coupling mechanisms between anion and proton translocation and of anion exchange-diffusion reactions in the inner membrane of rat-liver mitochondria is presented.

RESULTS AND DISCUSSION

Effect of pH on the Translocation of Anionic Substrates in mitochondria.

The effect of the pH of the suspending medium on the efflux of intramitochondrial Pi, malate and citrate is shown in Fig. 1. In the experiment of Fig. 1 A mitochondria were incubated at 25° in media of various pH's in the presence of oligomycin and rotenone. These inhibitors prevented substrate oxidation and energy supply.

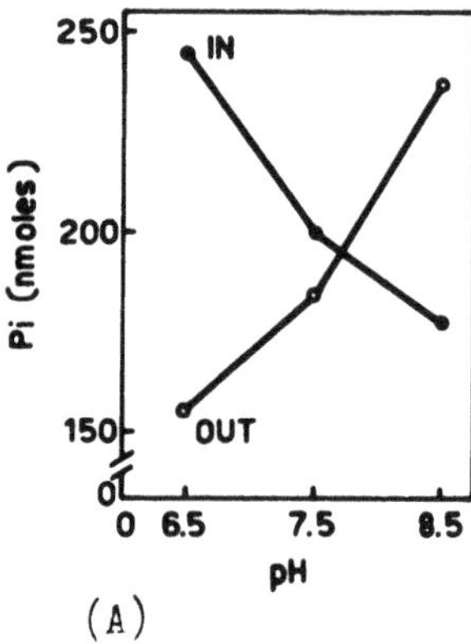

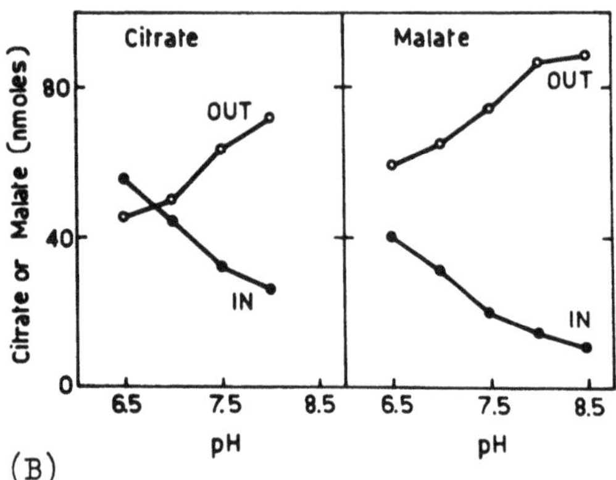

Fig. 1. Effect of the pH of the medium on the efflux of anionic substrates from mitochondria. Expt. A: mitochondria (11.7 mg protein) were suspended in a reaction medium containing: 15 mM KCl, 50 mM Tris-HCl, 5 mM $MgCl_2$, 2 mM EDTA, 1 mM arsenite, 3 µg rotenone and 15 µg oligomycin. Final volume 1.5 ml, temperature 25°. Expt. B: mitochondria were preincubated 10 min at 0° in 250 mM sucrose with 2 mM malate or 2 mM citrate, washed and resuspended at 0° in cold 250 mM sucrose containing 100 µM [^{14}C]malate or 100 µM [^{14}C] citrate (about 10^8 cpm). Loaded mitochondria (6.4 mg protein) were suspended in 1.5 ml of the reaction medium of Expt. A. For experimental details cf. Papa et al. (8). Substrate content in the matrix (●——●), in the supernatant (○——○).

The incubation was interrupted after 40 sec by rapidly centrifuging the mitochondria from the medium and the intra- and extra-mitochondrial level of Pi was determined. It can be seen that the increase of the pH from 6.5 to 8.5 promoted the efflux from the matrix of Pi. In the experiment of Fig. 1B mitochondria were preloaded with [^{14}C] malate or [^{14}C] citrate and incubated at 25° for 80 seconds in media of various pH's. The increase of the pH promoted also the efflux from the matrix of malate and citrate.

Further insight into the mechanism of this pH dependence was obtained by studying the initial phase of anion efflux. In the experiment of Fig. 2 mitochondria were first loaded with one single labelled substrate (Pi, malate, citrate or α-oxoglutarate), in the presence of oligomycin, rotenone and antimycin, and then exposed for a few seconds at 0° to a second incubation medium (free of

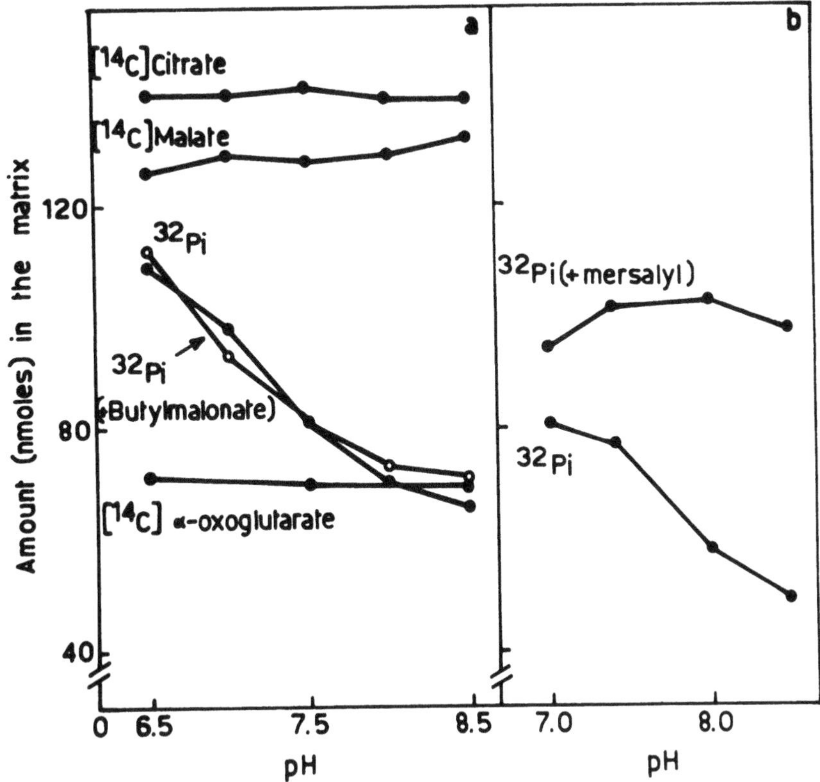

Fig. 2. Effect of a rapid exposure of mitochondria to media of various pH's on the intramitochondrial level of anionic substrates. Mitochondria (Expt. A, 7 mg protein; Expt. B 11 mg protein) preincubated at 25° were loaded with labelled anionic substrates and then centrifuged through a second incubation layer at 0°, containing the same components as the preincubation mixture, except anionic substrates, and, where indicated, 3 mM butylmalonate or 0.66 mM mersalyl. For experimental details see refs. 8,9.

anionic substrates) at various pH's. During the fast exposure to the second medium ^{32}Pi, accumulated during the first incubation, moved out of the mitochondria. On the contrary there was no significant efflux of [^{14}C] malate, [^{14}C] oxoglutarate or [^{14}C] citrate. It can be seen from Fig. 2 that as the pH of the second layer was increased, the efflux of Pi became greater. In contrast, changing the pH of the second layer from 6.5 to 8.5 had no effect

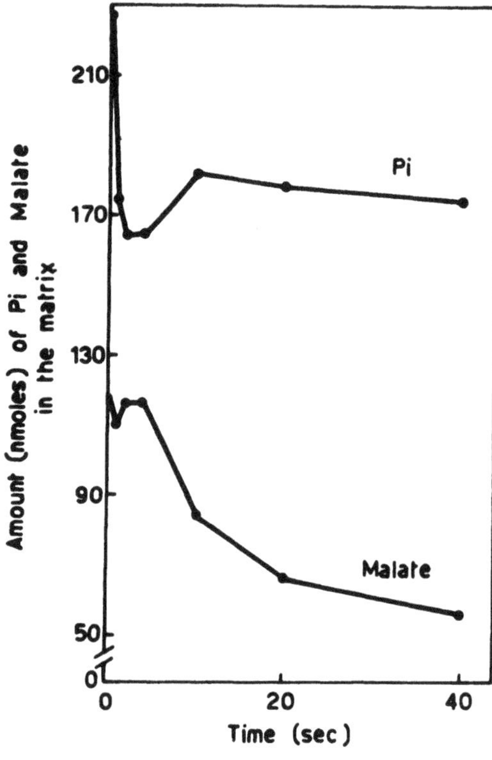

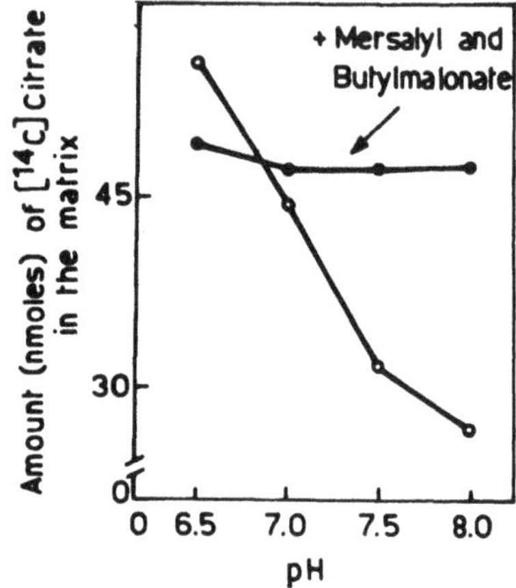

Fig. 4. Effect of butylmalonate plus mersalyl on the pH dependent efflux of citrate. Mitochondria loaded with $[^{14}C]$ citrate were resuspended in media at different pH's (6.4 mg protein). Where indicated 2 mM butylmalonate and 0.66 mM mersalyl were present from the beginning of the incubation.

Fig. 3. Time course of the efflux from mitochondria of endogenous Pi and malate. The experimental conditions are those described in the legend of Fig. 1 A - Mitochondrial protein 11.7 mg. At the times indicated 0.66 mM mersalyl was added and mitochondria rapidly centrifuged from the medium (8).

on the intramitochondrial level of malate, α-oxoglutarate and citrate under these conditions. The OH^- promoted efflux of Pi was blocked by mersalyl (Fig. 2,b) but not by butylmalonate (Fig. 2,a). Evidence has been presented that mersalyl and other sulphydryl-blocking reagents inhibit the transport of Pi in mitochondria (10-12).

The experiment of Fig. 3 gives a time-resolution of the events occurring during the efflux of endogenous Pi and malate from mitochondria at pH 8.5 and 25°. In the first 2 seconds of incubation a relatively rapid efflux of Pi occurred. During this interval there was no significant movement of malate. The Pi released into

TABLE I

Effect of Valinomycin and Butylmalonate on the Uptake of Anionic Substrates by Rat-liver Mitochondria

Mitochondria (6 mg protein) were incubated 1 min at 25° in the reaction medium described in Fig. 1, with the addition of 1 mM ^{32}Pi, 1 mM [^{14}C] malate, 1 mM [^{14}C] citrate or 1 mM [^{14}C] α-oxoglutarate; 0,5 μg antimycin, 0,3 mM N,N,N',N'-tetramethyl-p-phenilenediamine and 3 mM ascorbate. Where indicated: 0.1 μg valinomycin and 2 mM butylmalonate (see Fig. 1 and ref. 8).

Additions	Amount (nmoles) in matrix of			
	^{32}Pi	[^{14}C] Malate	[^{14}C] Citrate	[^{14}C] Oxoglutarate
None	157	158	125	103
Valinomycin	251	456	192	165
Butylmalonate	127	96	73	45
Butylmalonate + Valinomycin	319	75	56	46

the medium then started to move back <u>into</u> the mitochondria. At the same time malate started to move <u>out</u> of the mitochondria. In Fig. 4 the effect of mersalyl plus butylmalonate on the pH dependence of the efflux of citrate from mitochondria at 25° is shown. The incubation time was 80 sec. It can be seen that the stimulation of the efflux of citrate brought about by increasing the pH (see also Fig. 1) was completely abolished by mersalyl <u>plus</u> butylmalonate. It will be seen below that mersalyl and butylmalonate do not block, at the concentrations used here, the efflux of mitochondrial citrate induced by added malate.

It is known that valinomycin induces an energy-linked uptake of potassium by mitochondria (13,14) that is accompanied by H^+ extrusion and alkalinisation of the intramitochondrial space (15). The effect of valinomycin on the uptake of anionic substrates by rat-liver mitochondria is shown in Table I. ^{32}Pi, [^{14}C] citrate,

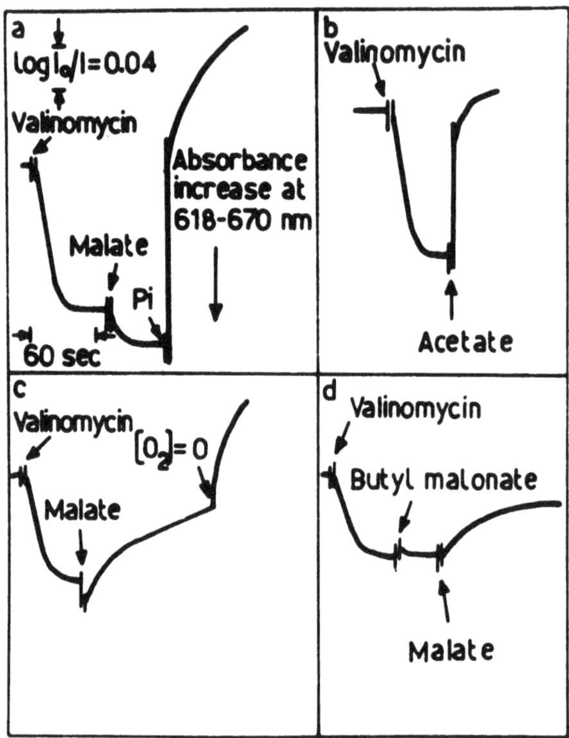

Fig. 5. Effect of malate, Pi and acetate on bromothymol blue absorbance changes in mitochondria. In Expts. a and b, mitochondria (9 mg protein) were preincubated for 4 min at 25° in the presence of 130 mM KCl, 20 mM Tris-HCl (pH 7.5), 5 mM glutamate, 0.2 mM ADP, 3 µM bromothymol blue, 10 mM glucose and 15 I.U. yeast hexokinase. After 2 min, 15 µg oligomycin were added followed 1 min later by 0.2 µg valinomycin. Expts. c and d, unpreincubated mitochondria. Additions: 3 mM malate, 3 mM Pi, 3 mM acetate and 1 mM butylmalonate (see ref. 9).

[^{14}C] malate, or [^{14}C]α-oxoglutarate were added to the mitochondrial suspension in the presence of oligomycin, rotenone and antimycin. Energy was supplied by the oxidation of N,N,N',N'-tetramethyl-p-phenylenediamine (plus ascorbate). The induction of K^+ uptake by the addition of valinomycin resulted in a marked stimulation of the uptake of Pi, malate, citrate, and α-oxoglutarate. A similar stimulation has been reported by several Authors (16-18). However the important point shown by the experiment of Table I is that butylmalonate abolished the stimulating effect of valinomycin on the accumulation of malate, α-oxoglutarate and citrate. On the other hand butylmalonate potentiated the stimulating effect of valinomycin on the uptake of Pi.

In the experiment of Fig. 5 bromothymol blue was included in the mitochondrial suspension, and its absorbance changes were recorded during the incubation. Evidence has been presented that bromothymol blue absorbancy responds to intramitochondrial pH changes (15,19) see however, refs. (20, 21). Mitochondria were suspended in a KCl medium and depleted of endogenous Pi by incubation with glutamate and ~P trap (Figs. 5a and 5b). The addition of valinomycin caused a fast and large increase of bromothymol blue absorbance, the kinetics of which were similar to those reported by Pressman et al. (14), for the appearance of H^+ in the medium. This absorbance increase indicates, at least in part, mitochondrial alkalinisation. The addition of malate gave a small further increase of absorbance, Pi, on the contrary, immediately reversed the absorbance increase; acetate gave an effect similar to that of Pi. When malate was added to the mitochondrial suspension without prior depletion of Pi (Figs. 5c and 5d), it caused some reversal of the valinomycin-induced absorbance increase of bromothymol blue. This effect of malate was partly inhibited by 1 mM butylmalonate; the inhibition could be increased by raising the concentration of butylmalonate. That Pi (or acetate) but not dicarboxylate anion can reverse mitochondrial alkalinisation, induced by cation accumulation, has also been shown by direct pH measurements on the dissolved mitochondrial pellet (22) or by the distribution of 5,5-dimethyl-2,4-oxazolidinedione (DMO) across the mitochondrial membrane (23). Furthermore Rossi and coworkers (24) have reported that the uptake of Pi lowers the H^+/K^+ ratio during K^+ uptake by mitochondria.

The results presented and other obtained in our laboratory (18,25) show, in agreement with Mitchell's proposal (6,7) (contrast ref. 26), that the distribution of substrate anions across the cristae membrane of mitochondria can be governed by a transmembrane pH difference. However whilst according to Mitchell (6,7) the various anions are translocated across the mitochondrial membrane by anion-OH^- exchange diffusion reactions, or anion H^+ symports it is clear, from the experiments shown, that the translocation of Pi, but not that of malate, citrate and α-oxoglutarate,

TABLE II

Exchange Diffusion Between Intramitochondrial ^{32}Pi and Extramitochondrial [^{14}C] Malate in Rat-Liver Mitochondria

2 mM ^{32}Pi was added to mitochondria (15.6 mg protein) depleted of endogenous anionic substrates. Where indicated, 2 mM [^{14}C] malate was present in Layer II. Δ^{32}Pi is the difference in the Pi content of the matrix in the presence and absence of malate and Δ[^{14}C] malate the difference in the malate content in the presence and absence of Pi (for details see ref. 8).

Additions to		Amount (nmoles) in matrix of	
Layer I	Layer II	^{32}Pi	[^{14}C] Malate
^{32}Pi	-	228	-
-	[^{14}C] Malate	-	68
^{32}Pi	[^{14}C] Malate	165	135
Δ^{32}Pi		- 63	
Δ[^{14}C] Malate			+ 67

can be directly coupled to an OH^- counterflux. With regard to this it is worth recalling that mitochondrial swelling caused by active uptake of Ca^{2+}, in the presence of succinate or other anionic respiratory substrates requires Pi (arsenate or acetate) (27).

The Exchange-Diffusion Reactions

The experiments that follow illustrate how a transmembrane ΔpH can govern the distribution across the inner mitochondrial membrane also of those anions whose translocation is not directly proton coupled. The existence in rat-liver mitochondria of a dicarboxylate (2,3), a tricarboxylate (2,3) and an α-oxoglutarate-translocator (28-31) has been proposed. According to Chappell et al.

(2-5) the dicarboxylate, tricarboxylate and α-oxoglutarate translocators can mediate anion-OH^- exchange-diffusion and dicarboxylate - Pi, tricarboxylate-dicarboxylate and α-oxoglutarate-dicarboxylate exchanges respectively.

In the experiment of Table II, mitochondria were preincubated with ADP, glucose and hexokinase in order to lower the content of endogenous Pi and respiratory substrates. ^{32}Pi, was then added in the presence of oligomycin, rotenone and TMPD (plus ascorbate). Mitochondria actively accumulated Pi. ^{32}Pi-loaded mitochondria were centrifuged through a second incubation layer free of Pi. The presence of malate in this layer promoted Pi efflux from mitochondria. Concomitantly malate was taken up by mitochondria.

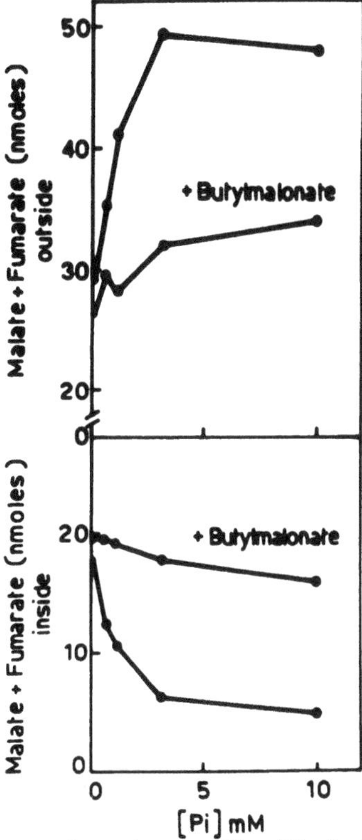

Fig. 6. Effect of butylmalonate on the Pi-driven efflux of endogenous malate from mitochondria. Mitochondria (5.9 mg protein) were suspended in 1.5 ml of the reaction mixture described in Fig. 1. Reaction temperature 20°, reaction time 1 min. Pi and 5 mM butylmalonate, were present from the beginning of the incubation (see Fig. 1 and ref. 8).

TABLE III

Exchange of Intramitochondrial α-Oxoglutarate with Extramitochondrial Malonate. Insensitivity of the Exchange to Butylmalonate

Rat-liver mitochondria (20.0 mg protein) were preincubated for 2 min with 125 mM KCl, 20 mM Tris-HCl (pH 7.5), 3 mM $MgCl_2$, 1 mM arsenite, 20 mM glucose and hexokinase. After 3 min (± glutamate) 4 µg rotenone, 2 µg antimycin and 20 µg oligomycin were added, and (where indicated) 1.5 mM [^{14}C] Malonate and 10 mM butylmalonate. After 2 min mitochondria were centrifuged from the suspending medium (see ref. 29).

Addition during		Amount (nmoles) in matrix of	
Preincubation	Incubation	α-Oxoglutarate	[^{14}C] Malonate
Glutamate, 10 mM	None	43	-
Glutamate	[^{14}C]Malonate, 1.5 mM	18	111
Glutamate	[^{14}C]Malnnate + Butylmalonate, 10 mM	20	85
None	[^{14}C]Malonate	0.1	93
None	[^{14}C]Malonate + Butylmalonate	0	56
-Butylmalonate	Δα-Oxoglutarate	-25	
	Δ[^{14}C] Malonate		+18
+Butylmalonate	Δα-Oxoglutarate	-23	
	Δ[^{14}C] Malonate		+29

Pi-loaded mitochondria took up twice as much [^{14}C] malate as unloaded mitochondria. In the unloaded mitochondria the residual amounts of endogenous anionic substrates must have been responsible for the uptake of [^{14}C] malate. The amount of ^{32}Pi driven out of the mitochondria by malate was approximately equal to the extra-uptake of malate. This experiment demonstrates the occurrence of an 1/1 exchange diffusion between malate and Pi. The experiment of Fig. 6 shows that the addition of Pi in the suspending medium promoted efflux from mitochondria of endogeneous malate. This effect of Pi was abolished by butylmalonate. Also the efflux of intramitochondrial Pi induced by malate was abolished by this inhibitor (not shown). These data show that butylmalonate inhibits the malate-Pi exchange-diffusion.

An experiment on the coupling between the flux of malonate and that of α-oxoglutarate across the mitochondrial membrane is presented in Table III. In the experiments of the first three lines glutamate was added to the mitochondria in the presence of Pi, ∼P acceptor and arsenite. The oxidation of glutamate generated α-oxoglutarate inside the mitochondria (28,29). Glutamate oxidation was then arrested with rotenone plus antimycin. On continuing the incubation much of the α-oxoglutarate produced was retained inside the mitochondria (line 1). The addition of malonate (line 2) promoted efflux of α-oxoglutarate. Concomitantly malonate was taken up by the mitochondria. In the absence of glutamate, no α-oxoglutarate was found inside the mitochondria, and much less malonate was taken up. It can be seen that the amount of α-oxoglutarate driven out of the mitochondria by malonate was practically equal to the extra uptake of malonate by α-oxoglutarate charged mitochondria. Thus there is a 1:1 exchange-diffusion of intramitochondrial α-oxoglutarate with extramitochondrial malonate. In Table IV the effect of butylmalonate is also shown. It inhibited the uptake of malonate, the extent of the inhibition being greater in uncharged mitochondria than in those charged with α-oxoglutarate. On the other hand, butylmalonate had no effect on the malonate-promoted efflux of α-oxoglutarate. The lower part of the table shows that practically the same amount of intramitochondrial α-oxoglutarate was exchanged with extramitochondrial malonate, in the absence and in the presence of butylmalonate. Thus α-oxoglutarate charged mitochondria took up malonate in two ways: in a butylmalonate-sensitive reaction (see Table I and fig.6) and in exchange-diffusion with α-oxoglutarate (butylmalonate-insensitive).

The occurrence in rat liver mitochondria of an analogous exchange-diffusion between malate and citrate is demonstrated by the experiment of Table IV. In Expt. A mitochondria were loaded with $[^{14}C]$ citrate at 0°, washed and then incubated for 80 sec at 25° in the presence of oligomycin and rotenone. Malate, added to the reaction mixture, was taken up by the mitochondria and, at the same time, promoted citrate efflux. The ratio between the nmoles of malate taken up and these of citrate driven out of the mitochondria was 1.9, The addition of mersalyl plus butylmalonate

TABLE IV

Citrate-Malate Exchange in Rat-liver Mitochondria

Expt. A: mitochondria were loaded with [^{14}C] citrate and subsequently suspended in the reaction medium at pH 6.5 (see Fig.1). Mitochondrial protein 6.0 mg, final volume 1.5 ml, reaction temperature 25°, reaction time 80 sec. Where indicate 1 mM malate, 0.66 mM mersalyl and 2 mM butylmalonate were added.

Expt. B: Mitochondria (3.3 mg protein) were preincubated at 20° in a medium containing 15 mM KCl, 50 mM Tris-HCl, pH 7.5, 2 mM EDTA, 5 mM $MgCl_2$, 1 mM ADP, 1.5 mM Pi, 1 mM arsenite. After 2 min 2 µg rotenone were added and 30 sec later mitochondria were loaded with the different concentrations of malate, 1 min later citrate + 5 mM butylmalonate were added. After 30 sec mitochondria were centrifuged from the medium (for details see ref. 8).

Expt. A

	Amount (nmoles) in matrix of				+ Mersalyl and Butylmalonate	
Additions	[^{14}C] Citrate	Malate	ΔMalate	ΔCitrate	ΔMalate	ΔCitrate
None	87.5	12.9				
Malate	42.0	100.0	+87.1	-45.5		
Mersalyl + Butylmalonate	92.0	10.2				
Malate + Mersalyl + Butylmalonate	53.7	39.5			+29.3	-38.3

Expt. B

$Malate_{out}$ mM (Added)	$Malate_{in}$ mM (Recovered)	$Citrate_{out}$ mM (Added)	$Citrate_{in}$ mM (Recovered)	ΔCitrate / ΔMalate
0.16	1.6	-	-	-
0.32	2.7	-	-	-
0.84	5.6	-	-	-
0.16	0.08	1.18	1.3	0.87
0.32	0.33	1.18	2.2	0.97
0.84	1.30	1.18	4.0	0.93

depressed severely malate uptake but had no significant effect on the malate driven efflux of citrate. In the presence of these inhibitors the ratio between the nmoles of malate taken up and those of citrate driven out of the mitochondria was 0.8. In Table IV,B an experiment on the exchange between intramitochondrial malate and extramitochondrial citrate is shown. Mitochondria were preincubated in the presence of Pi, phosphate acceptor and arsenite to deplete them of endogenous citrate. Mitochondria were then loaded with different amounts of malate in the presence of rotenone: 1 min after the addition of malate, citrate was added (together with butylmalonate to block malate-phosphate exchange). After 30 sec. mitochondria were centrifuged from the medium and the concentration of citrate and malate in the matrix was determined. It can be seen that citrate added to malate-loaded mitochondria was actively accumulated and, at the same time, promoted malate efflux. The ratio between the amount of citrate taken up and that of malate driven out ranged between 0.87 and 0.97. It is therefore evident that citrate, like α-oxoglutarate, is transported across the inner mitochondrial membrane by an exchange diffusion with malate, the stoicheiometry of the reaction being of one to one. This reaction, like the α-oxoglutarate-dicarboxylate exchange, is butylmalonate insensitive (see ref. 8).

The malate-citrate exchange presents however a specific problem. If the reaction consisted of an exchange of citrate^{3-} with malate^{2-} it would be electrogenic and should be stimulated by a compensatory flux of cation. On the other hand a neutral citrate^{2-} -malate^{2-} exchange would result in a translocation of H^+ across the membrane accompaning the citrate anion. In this case the reaction should be stimulated by promoting H^+ translocation in the direction opposite to that of the citrate flux. The experiment of Table V seems to distinguish between these two mechanisms. Mitochondria were loaded with [^{14}C] citrate. Butylmalonate and mersalyl were then added and the mitochondria transferred to a second incubation medium free of potassium. The presence of malate, in this medium, promoted an efflux of citrate and a stoicheiometric amount of malate was taken up by the mitochondria. The addition of nigericin, which in the absence of malate caused only a slight efflux of intramitochondrial citrate and intramitochondrial (endogenous) malate, caused, when added malate was present, a marked stimulation of the efflux of citrate and the extra-uptake of an approximately equal amount of malate. The stoichiometry of the citrate malate exchange remained approximately 1:1. Thus nigericin stimulates, under these

TABLE V

Effect of Nigericin on the Exchange of Intramitochondrial Citrate with Extramitochondrial Malate

Rat-liver mitochondria (6.0 mg protein), loaded with $[^{14}C]$ citrate were resuspended in the standard reaction mixture containing 0.66 mM mersalyl and 2 mM butylmalonate. For experimental details see Table IV and ref. 8.

Additions		nmoles in the matrix of	
		$[^{14}C]$ Citrate	Malate
None		92.0	10.2
Nigericin, 0.2 μg		71.5	8.2
Malate, 1 mM		53.7	54.5
Malate + Nigericin		13.0	81.0
- Nigericin	Δ-Citrate	-38.3	
	Δ-Malate		+44.3
+ Nigericin	Δ-Citrate	-58.5	
	Δ-Malate		+72.8

TABLE VI

Insensitivity of the Exchange of Intramitochondrial Citrate and Extramitochondrial Malate to Valinomycin

Mitochondrial protein 4.5 mg. Experimental conditions as in Table V (see also ref. 8).

Additions	Counts per min in the matrix of $[^{14}C]$ Citrate
None	85,600
Valinomycin, 0.2 μg	92,500
Malate, 1 mM	57,700
Malate + Valinomycin	51,100

conditions, the malate citrate exchange reaction. The data of Table VI show that, in contrast, valinomycin had no significant effect on the efflux of intramitochondrial citrate both in the absence and presence of external malate. Nigericin mediates an electroneutral $K^+ - H^+$ exchange-diffusion (14). In a potassium free medium it causes an influx of H^+ in exchange for mitochondrial K^+. Valinomycin on the other hand mediates under these conditions, an electrogenic efflux of K^+ (6,14,32). The fact that the citrate$_{in}$-malate$_{out}$ exchange is stimulated by nigericin, but not by valinomycin would indicate that this antiport consists of an electroneutral exchange of citrate^{2-} against malate^{2-}.

By exposing [^{14}C] malate loaded mitochondria to a second incubation layer containing increasing concentrations of the various anions that exchange with [^{14}C] malate (unlabelled malate, Pi, α-oxoglutarate and citrate) it was possible to see that the efflux of [^{14}C] malate, driven by the various counteranions, followed saturation kinetics. Lineweaver-Burk plots of these experiments gave for the exchange diffusion of malate with malate, Pi, α-oxoglutarate, and citrate the Vmax for malate efflux and the K_m for the various counteranions reported in Table VII. It may be noted that the K_m for the anions tested (ranged in the absence of other species) from 23 μM in the case of malate to 131 μM in the case of Pi. This relatively high affinity of anions for the various translocating systems may have important physiological implications.

The data presented show that the transport of dicarboxylate, α-oxoglutarate and tricarboxylate, across the inner membrane of rat-liver mitochondria occurs by specific exchange-diffusion reactions. The following exchange-diffusions are directly demonstrated: dicarboxylate with Pi, dicarboxylate with α-oxoglutarate and malate with tricarboxylate. The stoicheiometry of the Pi-dicarboxylate, dicarboxylate-α-oxoglutarate and malate-tricarboxylate exchange is one to one. Thus these reactions assure specific changes in the concentration of the various substrate anions in the two spaces, separated by the membrane, without affecting the total concentration of anions. There are evidences indicating that the Pi-OH^- (i)n Pi-dicarboxylate (ii) dicarboxylate α-oxoglutarate (iii) and malate tricarboxylate (iv) exchange-diffusion reactions are mediated by separate translocators: (I) reactions (ii)-(iv) exhibit a different specificity spectrum, towards the dicarboxylic acids (3). (II) The activity of the systems mediating

TABLE VII

Vmax of Malate Efflux from Rat-liver Mitochondria in Exchange-Diffusion with Various Anions and the K_m for the Counter Anion

Mitochondrial protein: 8,0 mg in Expts. 1 and 3; 7.0 mg in Expts. 2 and 4. Temperature 0°. For experimental details see ref. 33.

Expt.	Counter anion	Vmax (nmoles/15 sec)	K_m (µM)
1	Pi	45	131
2	Malate	67	23
3	Citrate	57	102
4	α-Oxoglutarate	51	56

Note that the $MgCl_2$ concentration was 1 mM and no account was taken of Mg^{2+} chelation by citrate in calculating the Km.

reactions (iii) and (iv) varies differently in mitochondria of different tissues (3). (III) At the concentrations rutinely used (25 nmoles/mg protein) mersalyl inhibits both reactions (i) and (ii) but has no effect on reactions (iii) and (iv) (12). Butylmalonate inhibits only reactions (ii) (12). N-ethylmaleimide inhibits only reaction (i) (34). (IV) Reactions (i)-(iv) have a different sensitivity to uncouplers of oxidative phosphorylation, and in the case of inhibition the kinetics of inhibition differs from system to system (33).

Although the four antiport reactions described appear to be mediated by four different translocators, each one can be coupled to the other through circulation of the substrates they share. The mechanism by which the different anion translocation reactions are coupled each other is shown in Fig. 7. This coupling allows the free energy, given by the concentration gradient of the various anions across the membrane to flow efficently from one system to the other. In this way a transmembrane ΔpH can goven the distribution across the inner mitochondrial membrane of all the anionic substrates shown in this scheme. Evidence has recently been

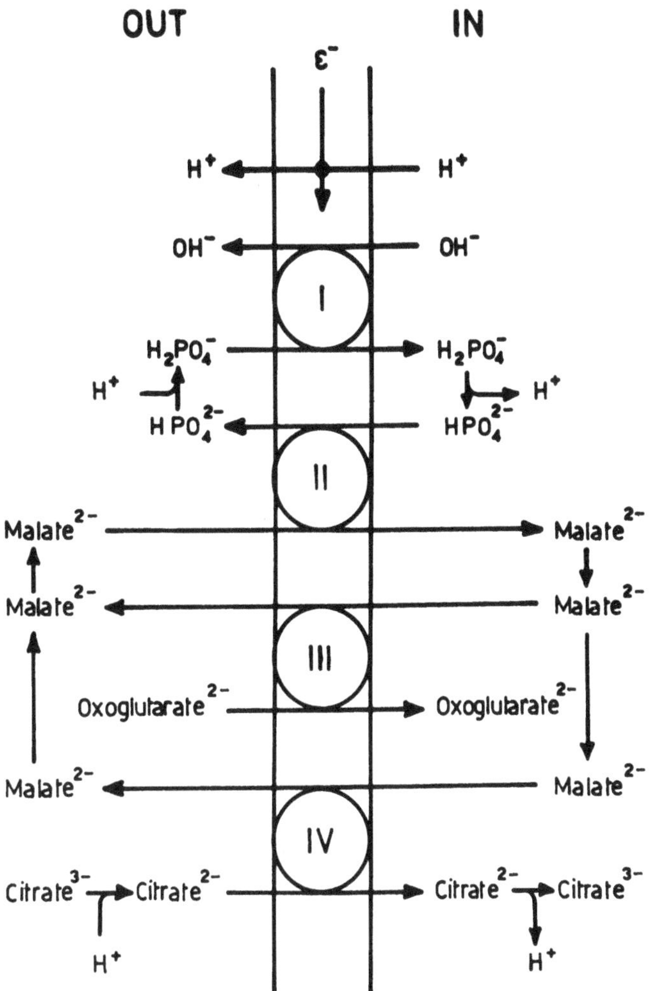

Fig. 7. Substrate Anion Translocation Across the Inner Membrane of Rat-liver Mitochondria.

obtained that also the movement of pyruvate across the inner mitochondrial membrane can be governed by transmembrane ΔpH (35). Further work is in progress to elucidate the exact nature of the reactions by which pyruvate moves across the mitochondrial membrane. It also follows from the mechanism shown in Fig. 7 that the free energy of the concentration gradient of the substrate anions can derive from that made available by the flux of electrons along the respiratory chain. It is known that in mitochondria electron flow can be coupled to ejection of protons in the surrounding medium. Various mechanisms have been proposed to explain this proton translocation (36-41). Thus, the energy made available by oxidoreductions can be, at least in part, converted to that of a proton gradient across the membrane. Since the translocation of Pi is protoncoupled, the free energy represented by the ΔpH can be converted into that of a Pi gradient and this in turn can flow into the other exchange-diffusion systems, all of these being connected. Papa, Quagliariello and Chance (42) have recently obtained evidence that the uptake of Pi by rat-liver mitochondria effects the energy pressure built up by the oxido-reductions of the respiratory chain. The addition of Pi to Pi-depleted rat-liver mitochondria, causes, under conditions of inhibition of oxidative phosphorylation a transient activation of electron flow. When reducing equivalents are fed to NAD this reaction is characterized by a rapid oxidation of flavoproteins and reduction of cytochrome c. In other experiments it has been found that in the presence of ascorbate as substrate and low concentrations of TMPD, Pi caused oxidation of

TABLE VIII

Correlation Between Pi-Uptake by Rat-liver Mitochondria and Reduction of Cytochrome c

Mitochondria (6.8 mg/ml) were preincubated 3 min in the standard reaction mixture containing in addition: 5 mM β-hydroxybutyrate, 0.2 mM ADP, 20 mM glucose, and 15 units of hexokinase. Oligomycin (10 µg) was then added, followed 60 sec later by 3 mM Pi. The other additions were made 30 sec before that of Pi. (From ref. 42).

Additions	^{32}Pi in the matrix (nmoles)	Pi-induced cytochrome c reduction, absorbance increase at 550 vs 540 mµ Log. Io/I
None	255	18×10^{-3}
DNP, 100 µM	108	0
Mersalyl 170 µM	16	0
Mersalyl+Cysteine 1 mM	257	18×10^{-3}

cytochrome c. The response of respiratory chain to Pi appears to be correlated with the uptake of Pi by mitochondria. The experiment of Table VIII shows that Pi, added in the presence of oligomycin, to Pi-depleted mitochondria, was actively accumulated by mitochondria. The uptake of Pi was severely inhibited by dinitrophenol. Mersalyl blocked almost completely the penetration of Pi, into the mitochondria. Both dinitrophenol and mersalyl suppressed the Pi-induced reduction of cytochrome c. Cysteine prevented completely the inhibitory effect of mersalyl on both Pi-penetration and cytochrome c reduction.

A link between electron flow and the influx of anionic substrates in mitochondria brings with itself the consequence that an increased burst of respiratory substrates can be automatically compensated by a replenishment of these substrates at the site where they are utilized.

AKNOWLEDGMENTS

This work was supported by a grant from "Consiglio Nazionale delle Ricerche" Centro di Studio sul Metabolismo Energetico e Mitocondri.

REFERENCES

1. Papa, S., Tager, J.M., Quagliariello, E. and Slater, E.C., "The Energy Level and Metabolic Control in Mitochondria", Adriatica Editrice, Bari (1969).
2. Chappell, J.B. and Haarhoff, K.N., in "Biochemistry of Mitochondria" (E.C. Slater, Z. Kaniuga and L. Wojtczak, eds.), p. 75, Academic Press and Polish Scientific Publishers, London and Warsaw (1967).
3. Chappell, J.B., Brit. Med. Bull., 24: 150 (1968).
4. Chappell, J.B., Biochem.J., 100: 43 (1966).
5. Chappell, J.B., Henderson, P.J.F., McGivan, J.D. and Robinson, B.H., in "The Interaction of Drugs and Subcellular Components on Animal Cell" (P.N. Campbell, ed.), p. 71, J.A. Churchill Ltd. London (1968).
6. Mitchell, P., "Chemiosmotic Coupling and Energy Transduction", Glynn Research Ltd., Bodmin, Kent (1968).

7. Mitchell, P., Adv. Enzymol., 29: 33 (1967).
8. Papa, S., Lofrumento, N.E., Quagliariello, E., Meijer, A.J. and Tager, J.M., J. Bioenergetics, (1970), in press.
9. Papa, S., Lofrumento, N.E., Loglisci, M. and Quagliariello,E., Biochim. Biophys. Acta, 189: 311 (1969).
10. Tyler, D.D., Biochem.J., 111: 665 (1969).
11. Fonyo, A., Biochem. Biophys. Res. Commun., 32: 624 (1968).
12. Meijer, A.J. and Tager, J.M., Biochim.Biophys. Acta, 189: 136 (1969).
13. Moore, C. and Pressman, B.C., Biochem.Biophys.Res. Commun., 15: 562 (1964).
14. Pressman, B.C., Harris, E.J., Jagger, W.S. and Johnson, J.H., Proc. Natl. Acad. Sci. U.S., 58: 1949 (1967).
15. Harris, E.J., FEBS Letters, 5: 50 (1969).
16. Lynn, W.S. and Brown, R.N., Biochim.Biophys. Acta, 110: 459 (1966).
17. Harris, E.J., in "Mitochondria Structure and Function", (L. Ernster and Z. Drahota, eds.), p.347, Academic Press, London (1969).
18. Palmieri, F. and Quagliariello, E., European J. Biochem., 8: 473 (1969).
19. Chance, B. and Mela, L., Proc. Natl. Acad. Sci. U.S., 55: 1243 (1966).
20. Mitchell, P., Moyle, J. and Smith, L., Eyropean J. Biochem., 4: 9 (1968).
21. Saris, N.E. and Seppala, A.J., European J. Biochem., 7: 267 (1969).
22. Fonyo, A., in "Electron Flow and Energy Conservation", (J.M. Tager, S. Papa, E. Quagliariello and E.C. Slater, eds.), Adriatica Editrice, Bari (1970) in press.
23. Addanki, S., Cahill, F. and Sotos, J.F., J. Biol. Chem., 243: 2337 (1968).
24. Rossi, C., Scarpa, A. and Azzone, G.F., Biochemistry, 6: 3902 (1967).
25. Palmieri, F., Quagliariello, E. and Klingenberg, M., Biochem. J., 116: 36P (1969).
26. Harris, E.J. and Pressman, B.C., Biochim.Biophys. Acta, 172: 66 (1969).
27. Chappell, J.B. and Crofts, A.R., Biochem.J., 95: 378 (1965).
28. De Haan, E.J. and Tager, J.M., Abstr. 3rd Meeting Feder. European Biochem. Soc., Warsaw (1966), p.159, Academic Press and Polish Scientific Publishers, London and Warsaw, (1966).

29. Papa, S., De Haan, E.J., Francavilla, A., Tager, J.M. and Quagliariello, E., Biochim.Biophys.Acta, 131: 14 (1967).
30. Robinson, B.H. and Chappell, J.B., Biochem.Biophys. Res. Commun., 28: 249 (1967).
31. Quagliariello, E., Papa, S., Meijer, A.J. and Tager, J.M., in "Mitochondria Structure and Function" (L. Ernster and Z. Drahota, eds.), p. 335, Academic Press, London (1969).
32. Mitchell, P. and Moyle, J., in "Biochemistry of Mitochondria" (E.C. Slater, Z. Kaniuga and L. Woitczak, eds.), p. 53, Academic Press and Polish Scientific Publishers, London and Warsaw (1967).
33. Lofrumento, N.E., Meijer, A.J., Tager, J.M., Papa, S. and Quagliariello, E., Biochim.Biophys. Acta, 197: 104 (1970).
34. Meijer, A.J., Groot, G.S.P. and Tager, J.M., FEBS Letters, 8: 41 (1970).
35. Papa, S. and Francavilla, A., unpublished observations.
36. Chappell, J.B. and Crofts, A.R., in "Regulation of Metabolic Processes in Mitochondria" (J.M. Tager, S. Papa, E. Quagliariello and E.C. Slater, eds.), p. 293, Elsevier, Amsterdam (1966).
37. Van Dam, K. and Kraanijenhof, K., in "The Energy Level and Metabolic Control in Mitochondria" (S. Papa, J.M. Tager, E. Quagliariello and E.C. Slater, eds.), p. 299, Adriatica Editrice, Bari (1969).
38. Mitchell, P., Biol. Rev., 41: 445 (1966).
39. Chance, B. and Mela, L., Nature, 212: 369 (1966).
40. Williams, R.J.P., in "Electron Flow and Energy Conservation" (J.M. Tager, S. Papa, E. Quagliariello and E.C. Slater, eds.) Adriatica Editrice, Bari (1970), in press.
41. Massari, S. and Azzone, G.F., European J. Biochem., 12: 310 (1970).
42. Papa, S., Quagliariello, E. and Chance, B., Biochemistry, 9: 1706 (1970).

EFFECT OF SALTS ON PROTON TRANSLOCATION ACROSS THE INNER MITOCHONDRIAL MEMBRANE

S. Papa, F. Guerrieri, M. Lorusso and E. Quagliariello

Department of Biochemistry, University of Bari (Bari)

Italy

Oxygen pulses in succinate-supplemented sonic submitochondrial particles, induce abrupt uptake of protons by the particles, followed by a stationary phase (1). With anaerobiosis there occurs an exponential release of protons which proceeds according to the first order equation (see Fig. 1). It is assumed that in the stationary state the rate of active proton influx is equal to that of proton efflux (2). The latter is given by the initial rate of proton efflux in anaerobiosis (2).

We have found that potassium salts stimulate the respiration-driven proton uptake in sonic particles (1). This effect was potentiated by valinomycin, indicating that, at least in the presence of this antibiotic, the stimulation of proton uptake was due to K^+ counterflux (cf. ref. 3). Since, however, the activity of K^+-salts varied with the anions used these must also be involved.

Table I summarizes the effects of a serie of salts on: the initial rate of proton uptake; the amount of proton taken up in the stationary state; the velocity constant of the efflux of protons in anaerobiosis; the rate of electron and proton flow in the steady state and the H^+/O ratio. Chloride, nitrate, iodide, thiocyanate and tetraphenylborate caused, with different efficacy, stimulation of the initial rate and the extent of respiration-driven proton uptake. This was accompanied by stimulation of the

TABLE I. EFFECT OF SALTS ON PROTON AND ELECTRON FLOW IN SONIC SUBMITOCHONDRIAL PARTICLES

The reaction mixture (1.5 ml, pH 7) contained 250 mM sucrose, 15 mM succinate, 0,4 mg purified catalase and 2.3 mg Mg-ATP sonic particles from beef-heart mitochondria (4). Temp. 25° : For other details see ref. 1. Na-TPB = Na-tetraphenylborate. The salts were added at concentrations giving approximately half-maximal stimulation of the initial rate of proton uptake.

Salts	Concen. (mM)	Stimulation Initial Rate H^+ Uptake %	Extent H^+ Uptake ngion/mg.Prot.	Velocity Constant H^+ Efflux sec−1	Stimulation Initial Rate H^+ Efflux %	Change Steady-State Respiration %	Steady State H^+/O
Control			5.00	0.590			0.50
KCl	15	37	6.69	0.252	15	+ 4	0.55
KNO_3	10	70	18.40	0.154	58	+ 3	0.77
KI	10	126	28.82	0.134	70	−22	1.09
KSCN	1	210	23.75	0.283	204	− 3	1.55
Na-TPB	0.05	155	15.90	0.154	115	+ 7	1.00
K_2SO_4	15	66	5.25	0.866	50	+27	0.59
K-Acetate	5	93	4.50	0.845	30	+30	0.50

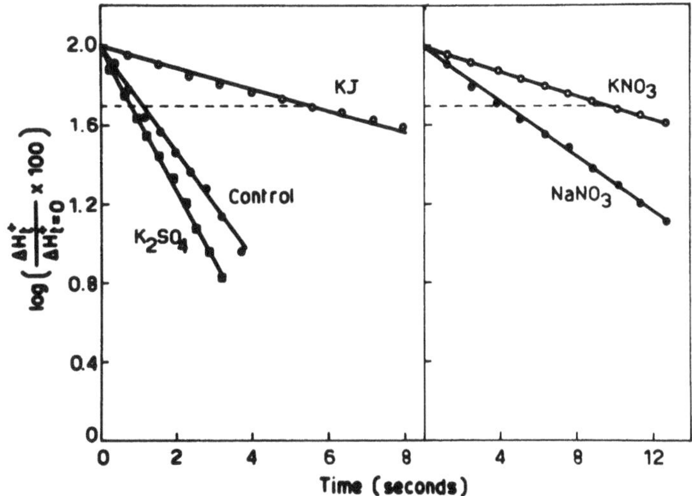

Fig. 1. Effect of salts on the velocity constant of anaerobic efflux of protons. For detail see Table I.

initial rate of the anaerobic efflux of protons but by a decrease of the velocity constant of this process (see Fig. 1). At the concentrations used these salts, except KI which caused inhibition, had practically no effect on respiration. This resulted in an increase of the steady-state H^+/O ratio. Sulfate and acetate increased the initial rate of proton uptake, the initial rate and the velocity constant of proton efflux (see Fig. 1). The extent of proton uptake was not significantly affected by these salts, respiration was stimulated.

In Fig. 1 a logarithmic plot of the anaerobic efflux of the amount of protons translocated in the stationary state is shown. It appears that whilst KI caused a considerable increase of the $t\frac{1}{2}$ of the proton efflux, K_2SO_4 lowered it.

The stimulation of the initial rate of proton uptake by the salts tested appears to be related to the permeability of artificial phospholipid membrane (5-7) and mitochondrial membrane (8) to the anionic species. Nitrate, iodide, thiocyanate and tetraphenylborate appeared to increase the steady-state turnover of protons, without stimulating respiration. This finding can be explained in terms of a proton pump (9,10) of electrogenic nature (9, see however refs. 11-13). In this case the thermodynamic potential difference of protons across the membrane consists of

TABLE II. EFFECT OF KNO_3 AND $NaNO_3$ ON PROTON TRANSLOCATION IN SONIC SUBMITOCHONDRIAL PARTICLES

Temp. 25°; pH, 7. For other details see legend to Table I.

Salts	Stimulation Initial Rate H^+ Uptake %	Extent H^+ Uptake ngion.mg prot.	Stimulation Initial Rate H^+ Efflux %	Velocity Constant H^+ Efflux sec^{-1}
KNO_3, 20 mM	50	41.32	63	0.087
$NaNO_3$, 20 mM	47	20.04	113	0.154

a chemical (Δ pH) and an electrical ($\Delta\psi$) component. Both exert a back-pressure on the proton uptake and drive proton efflux. A salt, which dissipates the $\Delta\psi$ through distribution of the anion in the electric field, stimulates proton uptake and converts the proton efflux driven by the electric field (probably not seen by the electrode, due to the short time life of the electric field) (2) into a pure diffusion. Thus the increase of the proton turnover would be, in large part, only apparent. This explanation is also supported by the fact that these salts caused a decrease of the velocity constant of proton efflux. When the $\Delta\psi$ is dissipated the diffusion of protons, in anaerobiosis, becomes electrogenic and depends upon the compensatory movement of other ions. In favour of an electrogenic proton pump is also the fact that respiration causes accumulation of the anions by the particles (7). We have found (14) that respiration caused an accumulation of [^{14}C] thiocyanate by the particles which was partly depressed by valinomycin and completely suppressed by uncouplers or lytic detergents. Sulfate and acetate increased the velocity constant of proton efflux and stimulated respiration. The increased rate of proton efflux is possibly due to back-diffusion of the acid. This results in a net increase of energy-expending proton turnover across the membrane. Such mechanism would be similar to that predicted by the chemiosmotic hypothesis for the action of uncouplers of oxidative phosphorylation (9).

Table II shows that $NaNO_3$ was more effective than KNO_3 in stimulating proton efflux. $NaNO_3$ gave a higher velocity constant of proton efflux (see Fig. 1), but a smaller extent of proton uptake than KNO_3. This suggests that the pH driven proton efflux can be, at least in part, mediated by K^+-H^+ and more actively, by Na^+-H^+ antiport systems (see ref.9). The fact that Na^+ gives a higher rate of proton efflux but a lower rate of proton uptake than K^+, would indicate that the cation-proton antiport system do not take part in the active influx of protons (contrast ref. 13).

REFERENCES

1. Papa, S., Guerrieri, F., Rossi Bernardi, L. and Tager, J.M., Biochim.Biophys. Acta, 197: 100 (1970).
2. Rumberg, B., Reinwald, R., Schroder, H. and Siggel, U., Naturwissenschaften, 55: 77 (1968).
3. Jackson, J.B., Crofts, A.R. and Von Stendingk, L.V., European J. Biochem., 6: 41 (1968).
4. Low, H. and Vallin, I., Biochim.Biophys.Acta, 69: 361 (1963).
5. Lauger, P., Lesslauer, W., Marti, E. and Richter, J., Biochim.Biophys.Acta, 135: 20 (1967).
6. Liberman, E.A. and Topaly, V.P., Biochim.Biophys. Acta, 163: 125 (1969).
7. Liberman, E.A., Topaly, V.P., Tsofina, L.M., Jasaitis, A.A. and Skulachev, V.P., Nature 222: 1076 (1969).
8. Mitchell, P. and Moyle, J., European J. Biochem., 9: 149 (1969).
9. Mitchell, P., "Chemiosmotic Coupling and Energy Transduction", Glynn Research Ltd., Bodmin Kent (1968).
10. Chappel, J.B. and Haarhoff, K.N., in "Biochemistry of Mitochondria" (E.C. Slater, Z. Kaniuga and L. Wojtczak, eds.), p. 75, Academic Press, New York (1967).
11. Chance, B. and Mela, L., Proc. Natl.Acad. Sci. U.S., 55: 1243 (1966).
12. Pressman, B.C., Proc. Natl. Acad. Sci. U.S., 53: 1076 (1965).
13. Massari, S. and Azzone, G.F., Europ.J. Biochem., 12: 310 (1970).
14. Papa, S., and Lorusso, M., unpublished observations.

Ca^{++} AND MITOCHONDRIAL MEMBRANES: EVIDENCE FOR SPECIFIC ENZYMIC CARRIERS

E. Carafoli, P. Gazzotti, C.S. Rossi and R. Tiozzo
Institute of General Pathology, University of Modena
and University College, Nairobi (Kenya)

The field of Ca^{++} transport in mitochondria has become rather complicated in recent times. In addition to the energy-linked transport, which has been studied intensively in the last seven or eight years (1), other Ca^{++} binding processes have been described. Azzone and his group (2-3) have reported the binding of rather large amounts of Ca^{++} by liver mitochondria in the absence of metabolism; part of the binding takes place at the surface of the mitochondrion, and is thought to reflect the reaction of Ca^{++} with the phospholipids of the mitochondrial membrane, another part probably occurs in the intramitochondrial space, and is linked to the opposite translocation of a stoichiometric amount of intramitochondrial K^{+}: the efflux of K^{+} is responsible for the uptake of Ca^{++}. More recently, Reynafarje and Lehninger (4) have described the so called "high-affinity" binding of very small amounts of Ca^{++} by isolated mitochondria, and have suggested that the phenomenon reflects the existence of a carrier molecule specific for Ca^{++} in the mitochondrial membrane.

In this paper, the most significant properties of the various Ca^{++} binding processes will be briefly described; emphasis will be put on those which are relevant to the main purpose of this presentation, which is that of surveying the evidence for the existence of a Ca^{++} carrier, and of discussing its role in the process by which isolated mitochondrial actively translocate Ca^{++} from the external medium.

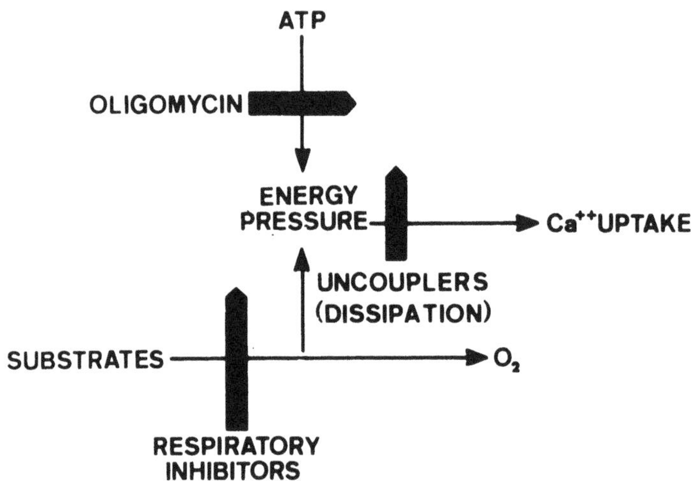

Fig. 1. Energetics of Ca^{++} transport in mitochondria.

Figure 1 shows that the energy-linked Ca^{++} transport requires the expenditure of the energy pressure maintained in mitochondria by either respiration or ATP hydrolysis. The respiration-driven process is inhibited by respiratory chain inhibitors and not by oligomycin, whereas the opposite is true for the ATP-driven process. Both systems are inhibited by uncouplers. Although the nature of the energy-pressure is still debated, it is important to mention that it is generally accepted that there are two levels of energy conservation, only the second involving the partecipation of phosphate. It is also generally accepted that uncouplers discharge the non-phosphorylated high-energy level, whereas oligomycin acts at the phosphorylated level. The effects of the inhibitors clearly indicate that the energy-linked uptake of Ca^{++} does not involve the discharge of phosphorylated levels of energy conservation; it has been shown long ago that energy-linked Ca^{++} uptake can proceed in the complete absence of inorganic phosphate, up to an uptake level of about 100–150 nMoles of Ca^{++} per mg of mitochondrial protein. This amount probably represents the saturation of fixed anionic charges in the mitochondrial membranes; however, when phosphate is present, much more Ca^{++} can be taken up, since phosphate penetrates into the mitochondria and removes Ca^{++} as insoluble salt. The formation of Ca^{++}-phosphate precipitates inside the mitochondria has been documented (5).

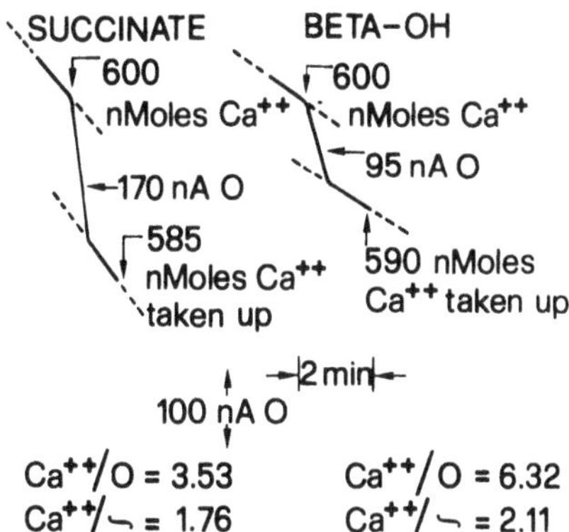

Fig. 2. Stoichiometry of oxygen consumption and Ca^{++} uptake in mitochondria. Oxygen is measured polarographically, and Ca^{++} isotopically. The medium contains 80 mM NaCl 10 mM Tris-Cl, pH 7.2, 10 mM respiratory substrate, and 2.5 mg mitochondrial protein per ml. Temp., 25°.

When supported by the respiratory chain, as is usually the case, the energy-linked uptake of Ca^{++} is associated with a reversible stimulation of respiration, which is shown in Fig. 2. The amounts of "extra" oxygen consumed, and of Ca^{++} taken up, are stoichiometrically related: about 2 Ca^{++} atoms are taken up as a couple of electrons is transferred across each one of the 3 energy-conserving sites of the respiratory chain. Parallel to the stimulation of respiration, there is also a reversible shift of the redox state of the respiratory carriers towards a more oxidized state (6). The transition of cytochrome b is perhaps the most sensitive index of the energy-linked reaction of Ca^{++} with mitochondria, and has been widely used to calculate the affinity of the energy-linked system for Ca^{++}. The K_m values most frequently found in the literature are around $5x10^{-5}$M. However, we have recently used Ca^{++}-EGTA(+) buffers to obtain low, but stabilized, Ca^{++} concentrations, and have used them to induce the shifts of the redox state of cytochrome b; we have thus been able to obtain a K_m 50 times lower, about 10^{-6}M (7). Another aspect of the energy-linked uptake of Ca^{++} is illustrated in Fig. 3; the movements of Ca^{++} are coupled to the opposite translocation of protons; this phenomenon is very useful, since it permits the continuous monitoring of the Ca^{++} uptake with a sensible pH meter. The amount of H^+

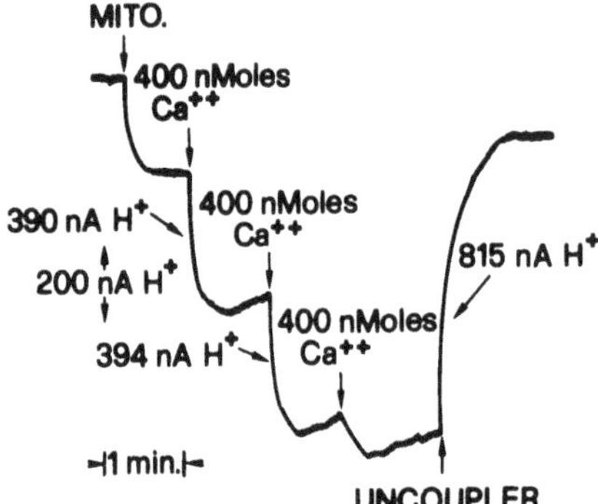

Fig. 3. Ca^{++} uptake and H^{+} ejection in mitochondria. Conditions as in Fig. 2, except for the concentration of Tris-Cl (5 mM).

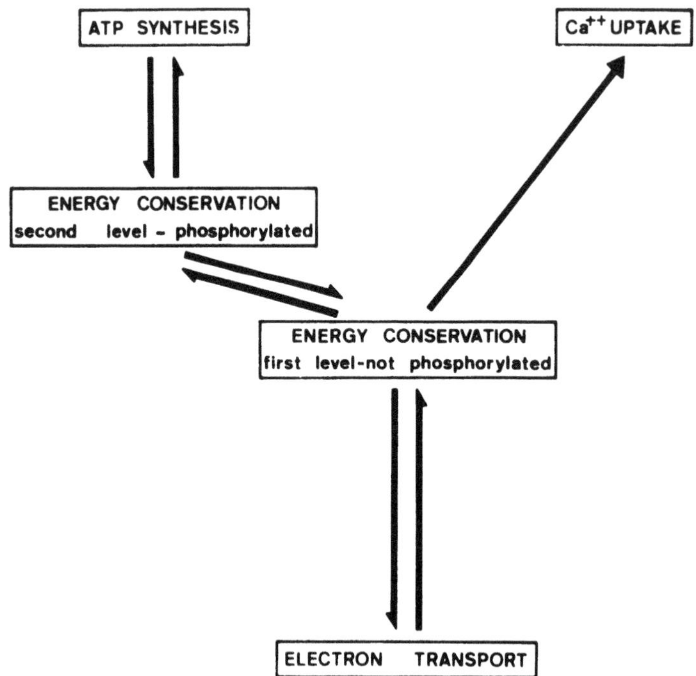

Fig. 4. Energy-linked Ca^{++} transport, and phosphorylation of ADP, as alternative processes.

ejected depends on a number of conditions, but in the absence of permeant anions the ratio H^+ ejected/Ca^{++} taken up is generally about 1. As seen in the Figure, also the maintenance of the accumulated Ca^{++} in mitochondria requires energy. Uncouplers indeed discharge Ca^{++}, and induce the reabsorption of a stoichiometric amount of H^+.

The energy-linked translocation process probably has physiological implications. The effect of the inhibitors and the reactivity of the respiratory chain, which clearly indicate the utilization of the energy-coupling sequence, together with the fact that mitochondria remain functionally intact during the process, suggest that Ca^{++} uptake and the phcsphorylation of ADP are alternative processes, as visualized in Fig. 4. Mitochondria can thus be considered organelles with a dual function: the production of ATP, and the regulation of the concentration of Ca^{++} in the cytosol (8).

As mentioned, mitochondria can bind considerable amounts of Ca^{++} in the absence of metabolism; in the absence of K^+ efflux, a maximum of about 50 nMoles of Ca^{++} can be bound in this process, with a rather low affinity: the apparent K_m is of the order of 10^{-4}M. Precise indications that the process involves the phospholipids of the mitochondrial membranes came from the observation by Scarpa and Azzi (9) that reagents capable of displacing Ca^{++} from phospholipid micelles, like the local anaesthetic butacaine, inhibited the process. Later, it was found that the binding was suppressed when mitochondrial phospholipids were extracted with acetone, and was restored specifically by the addition of phosphatidylethanolamine (10). Although the involvment of phospholipids in this type of binding seems well established, we have some reservations on the specific role of phosphatidyl-ethanolamine; this point will be considered in detail below. The process in severely inhibited also by monovalent cations, and it is of considerable interest that under conditions in which this type of binding is inhibited, for example in the presence of butacaine, the energy-linked translocation of Ca^{++} is accelerated (11).

In the absence of metabolism, and in the presence of high concentrations of NaCl to decrease the extent of phospholipid binding, uptake of Ca^{++} can be induced by the addition of valinomycin, which discharges the endogenous K^+. The phenomenon is shown in Fig. 5, which is redrawn from a Figure recently published by

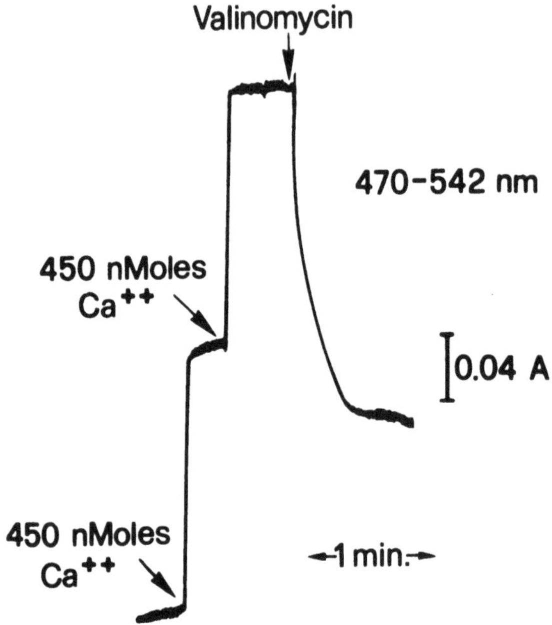

Fig. 5. K^+-induced Ca^{++} uptake (after Scarpa and Azzone, 3).

Scarpa and Azzone (3), and in which Ca^{++} was monitored continously with the indicator murexide in a dual wavelenght spectrophotometer. The extent of the uptake is determined by the K^+ store in mitochondria, and the permeabilization of the mitochondrial membrane with valinomycin is absolutely required. Since the incubation medium does not contain K^+, the K^+ concentration gradient across the mitochondrial membrane is discharged upon addition of valinomycin, to be replaced by a Ca^{++} concentration gradient. Scarpa and Azzone (3) have found that the K^+-dependent Ca^{++} uptake is abolished when the extramitochondrial concentration of K^+ is raised above 13 mM, and this confirms that the driving force for the uptake of Ca^{++} is indeed the K^+ concentration gradient. Scarpa and Azzone (3) have found that the K^+-driven Ca^{++} uptake is inhibited by uncouplers of oxidative phosphorylation, exactly as the energy-linked Ca^{++} uptake. Very interestingly, they have found that the process is inhibited also by La^{+++} ; as will be seen below, La^{+++} is considered a specific inhibitor of the hypothetical Ca^{++} carrier; the Ca^{++} carrier is thus probably involved also in the K^+ driven Ca^{++} translocation that takes place in the absence of metabolism.

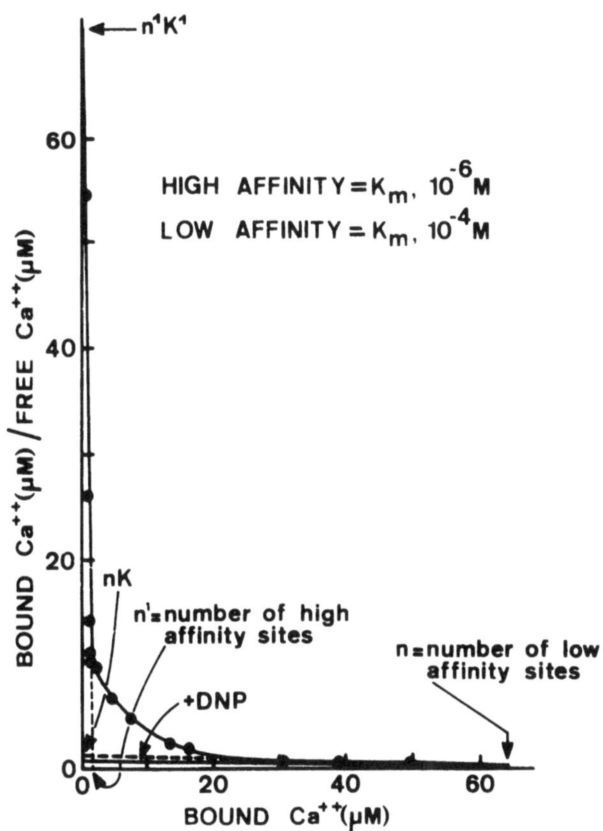

Fig. 6. Statchard plot of Ca^{++} binding by mitochondria. Reaction medium: 250 mM sucrose, 10 mM Tris-Cl, pH 7.2, 1 μM rotenone, 0.2 μg/mg protein Antimycin A. Mitochondrial concentration, 5 mg protein, in a final volume of 2 ml. Temp., 0°. Incubation time, 1 min.; Ca^{++} was measured isotopically.

In 1969, Reynafarje and Lenhinger (4), during a study of the metabolism-independent binding of Ca^{++}, added minimal amounts of Ca^{++} to rat liver mitochondria, and discovered that they were bound with great affinity. When the Ca^{++} to protein ratios were higher, the affinity was lower, as expected from the literature. They made the important observation that the high-affinity binding that took place at very low Ca^{++}/protein ratios was completely abolished by uncouplers, a fact that suggested the involvment of the high-affinity, Ca^{++} binding reaction in the process of energy-linked Ca^{++} uptake. A plot of their data according to Scatchard, shown in Fig. 6, revealed a biphasic curve, which indicated 2 classes of Ca^{++} binding sites. Extrapolations from the plots showed that one class had about 50 binding sites per mg of mitochondrial protein, with a rather low affinity for Ca^{++}; the apparent K_m was around 10^{-4}M. Clearly, these sites corresponded to the phospholipid sites responsible for the metabolism-independent, surface binding studied by Azzone and his group (2). The other class had only about one site per mg of protein, and its affinity for Ca^{++} was very high: the apparent K_m in some cases was 10^{-7} M. As seen in Fig. 6, the binding of Ca^{++} to the high affinity sites was completely abolished by DNP. High affinity sites were described also by Mela (12) and Mela and Chance (13), in a study of the early phases of energy-linked Ca^{++} uptake. They made the important observation that La^{+++} at low concentrations specifically inhibited this type of binding.

Both our group, and the group of Chance, have speculated that the high affinity reaction indicates the presence of a carrier molecule capable of transferring Ca^{++} across the mitochondrial membrane, and presumably active in the energy-linked uptake of Ca^{++}. The reason for this suggestion cannot be discussed in detail here; however, the key points of the rationale can be mentioned. They were the saturability of the energy-linked uptake, which differentiates it from the simple physical diffusion, the affinity of the high-affinity sites for Ca^{++}, and their specificity for it (see below). Also, we had shown previously that under some conditions Ca^{++} can oscillate back and forth across the mitochondrial membrane without changes in the rate of electron transport (14), suggesting that under certain conditions Ca^{++} can enter and leave mitochondria via a specific carrier, independent from electron transport, and capable of operating in either direction according to concentration gradients; energy was necessary to move Ca^{++} against a gradient. Mela and Chance (12,13) have based their suggestion of a carrier on the inhibition by La^{+++}; they consider it a specific inhibitor of the carrier, as are for instance butyl-malonate, atractyloside, and mersalyl, for their respective mitochondrial permeases. We have found that La^{+++}, at the same concentrations which are active on the early phases of energy-linked Ca^{++} uptake, very effectively inhibits high-affinity binding. This is shown in Fig. 7. The Scatchard plot in the presence of La^{+++} is rectilinear; only one leg of the plot is left, corresponding to the low-affinity sites; on the other hand, butacaine does not inhibit high-affinity binding. It must be mentioned here that very recently Scarpa and Azzone (3) have reported no inhibition of high affinity binding by La^{+++}. The reasons for the discrepancy are not clear: however, the inhibition by La^{+++} is a clear-cut phenomenon, very reproducible, and has been observed also in other Laboratories (15). Possibly, the endogenous phosphate of mitochondria may be the factor involved in the experiments of Scarpa and Azzone (3); it could mask the effect of La^{+++}, since phosphate complexes La^{+++} very efficiently. There is an aspect of the La^{+++} problem which is puzzling. A specific inhibitor should block the operation of a carrier in either direction: this is for example the case for mersalyl, which blocks both the entrance and the exit of phosphate in mitochondria (16). In the presence of La^{+++}, both the entrance and the exit of Ca^{++} should thus be inhibited. However, we have recently found that La^{+++} inhibits only negligibly the DNP-induced release of accumulated Ca^{++}. This is seen in Fig. 8, in which the movements of Ca^{++} were measured by following the opposite movements of H^+. La^{+++} slows down slightly the rate of Ca^{++} release, but it is clear that under the conditions used in this experiment the hypothetical carrier can still transport Ca^{++} out of the mitochondria, despite the presence of La^{+++}.

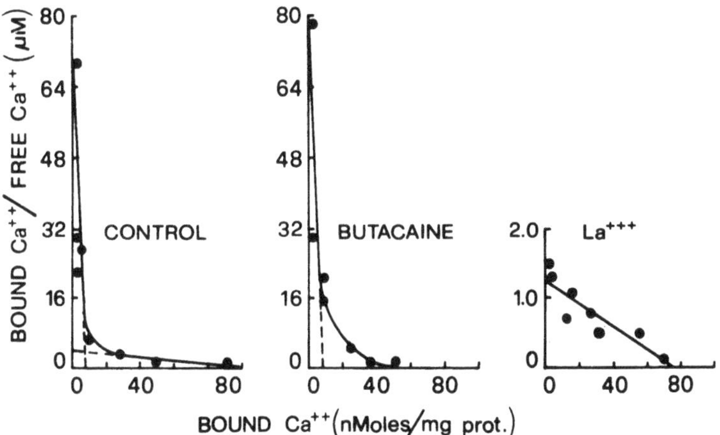

Fig. 7. Inhibition of high-affinity binding by La^{+++}. Conditions as in Fig. 6. Concentration of La^{+++}, 1 μM, of butacaine 0.1 mM.

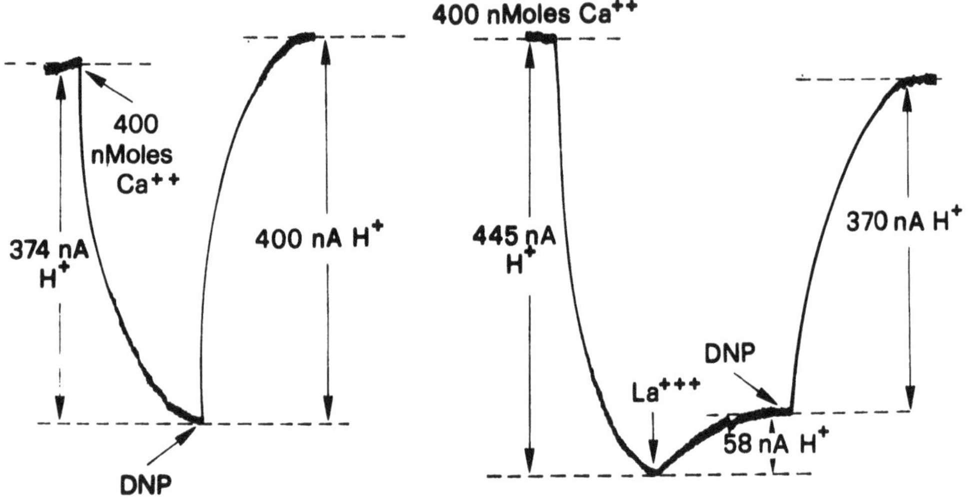

Fig. 8. Effect of La^{+++} on the release of Ca^{++} from mitochondria. Conditions as in Fig. 3.

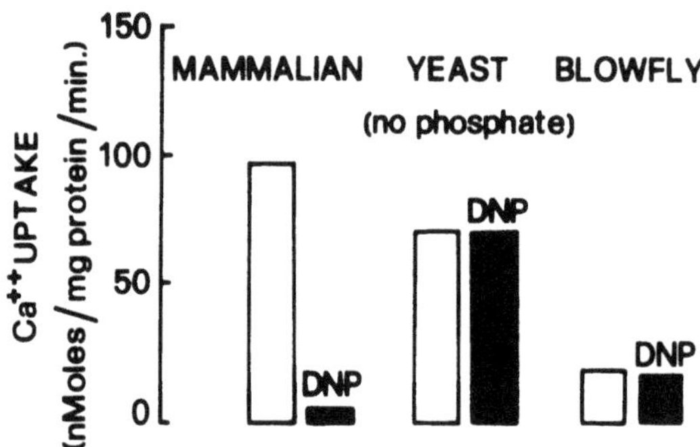

Fig. 9. Energy-linked Ca^{++} transport in different mitochondria. Conditions of incubation as in Fig. 2 (mammalian mitochondria), Fig. 10 (blowfly), and as described by Carafoli et al. (21) (yeast). Time, 1 min. DNP, 0.1 mM.

It must also be mentioned that Reynafarje and Lehninger (4) have found that the high-affinity sites are rather specific for Ca^{++}. Specificity, high affinity, saturability, possibility of functioning independently from the respiratory chain, and sensitivity to a specific inhibitor, La^{+++}, are thus the main properties of the high-affinity binding sites; undoubtedly, these properties suggest a Ca^{++} carrier. We have obtained additional evidence for a specific Ca^{++} carrier in the mitochondrial membrane from an entirely different line. Since it is known that other mitochondrial permeases are absent from certain mitochondrial types, probably due to genetic determination (17), we have examined the distribution of the high-affinity binding in mitochondria from different sources, looking for mitochondria where the reaction was absent. We have surveyed the energy-linked uptake and the phospholipid binding as well, to establish whether energy-linked Ca^{++} transport is an universal property of mitochondria, as is the phosphorylation of ADP, and to gain information on the relationships among the different mechanisms of Ca^{++} binding. Our survey has shown that energy-linked Ca^{++} transport is present in all mammalian mitochondria tested so far (18-21); this is shown in Fig. 9, in which the inhibition of Ca^{++} transport in respiring mitochondria by DNP has been used to establish the presence of energy-linked transport. Clearly, under these conditions the transport was apparently absent from yeast and blowfly flight muscle mitochondria; at a first glance, this finding would suggest that energy-linked transport is possibly a property of mammalian mitochondria only. However, under appropriate conditions it is possible to demonstrate a certain amount of energy-linked transport in yeast mitochondria; to this purpose, amounts of Ca^{++} 10-50 times higher than those normally used for mammalian mitochondria must be added. As for blowfly flight muscle mitochondria, they take up large amounts of Ca^{++} in a reaction which is sensitive to respiratory chain inhibitors and to uncouplers, but inorganic phosphate is mandatorily required for the process: as shown in Fig. 10, the inhibitor of the phosphate carrier mersalyl blocks the uptake of Ca^{++}. Under these conditions, very little uptake of Ca^{++} takes place when phosphate is replaced by acetate; however, we have found that when the amount of Ca^{++} added is decreased to 20-50 nMoles per mg of mitochondrial protein, acetate becomes almost as efficient as phosphate in supporting the uptake of Ca^{++}. Since it is known that the penetration of acetate into mitochondria is a passive phenomenon, it must be concluded that also in these mitochondria the species actively

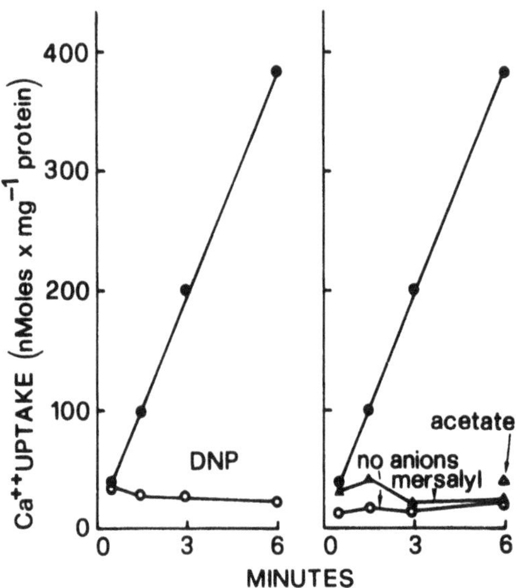

Fig. 10. Effect of phosphate on the uptake of Ca^{++} by blowfly flight muscle mitochondria. The reaction medium contains 250 mM sucrose, 20 mM inorganic phosphate, 10 mM pyruvate, 5 mM proline, 5 mM $MgCl_2$, 0.16 mM Ca^{++}, and 5 mg of mitochondrial protein. Concentration of DNP is 0.1 mM, of mersalyl, 0.1 mM, of acetate, 20 mM. Volume, 8 ml. Temp. 25°. Ca^{++} was measured isotopically.

transported is Ca^{++}. However, the mandatory requirement for anion translocation is peculiar; blowfly flight muscle mitochondria are apparently unable to couple the active uptake of Ca^{++} to the antiport of H^+ or other cations.

Table I shows that high-affinity binding is present in mammalian mitochondria; the number of sites varies from 0.2 to 11.0 per mg of mitochondrial protein, and their affinity for Ca^{++} is invariably very high. On the other hand, the high affinity sites are absent from blowfly flight muscle and yeast mitochondria. The absence of the sites from blowfly flight muscle mitochondria parallels the absence of other mitochondrial carriers from the same mitochondria, and is in line with the idea of a specific, genetically determined, carrier compound. Table I also shows that the low-affinity sites are present in mammalian and yeast mitochondria, but very poorly represented in blowfly flight muscle

TABLE I

High and Low-Affinity Ca^{++} Binding in Different Mitochondria

Mitochondria from	HIGH AFFINITY number of sites per mg protein	K_m (μM)	LOW AFFINITY number of sites per mg protein	K_m (μM)
Mammalian tissues	0.2–11	0.4–9.7	14–66	12–129
Yeast	none	=	39–50	11–39
Blowfly flight muscle	none	=	2	=

mitochondria. This is intersting, since the low affinity sites reflect the binding of Ca^{++} to mitochondrial phospholipids; blowfly flight muscle mitochondria could thus represent an useful tool to evaluate the specific phospholipid component responsible for the metabolism-independent binding. We have examined the phospholipid composition of blowfly flight muscle mitochondria and have indeed found that it differs considerably from that of liver, heart, or kidney mitochondria, as seen in Table II. The most striking difference concerns phosphatidyl-ethanolamine, which is far more abundant than in mammalian mitochondria; this is surprising, in view of the conclusion by Scarpa and Azzone (11) that phosphatidylethanolamine is the phospholipid specifically responsible for the metabolism-independent Ca^{++} binding by liver mitochondria. It is possible that what is true of liver mitochondria is not of blowfly muscle; however, we have recently obtained another piece of evidence which cast some doubts on the specific role of phosphatidylethanolamine in the metabolism-independent binding of Ca^{++}. Table III shows an experiment on mitochondria aged for two days at 2°. The changes in the content of phosphatidylcholine during aging are negligible, but phosphatidylethanolamine drops very markedly (about 40%). Yet, the number of low affinity sites changes only from 82 to 75 per mg of mitochondrial protein.

TABLE II

Phospholipid content and low-affinity Ca^{++} binding sites in aged mitochondria

	Phosphatidyl-choline µg lipid P/mgN	Phosphatidyl-ethanolamine µg lipid P/mgN	Number of low-affinity Ca^{++} binding sites per mg mitochondrial protein*
Control mitochondria	21.2	16.5	82
Mitochondria aged 48 hours	19.9	9.3	75

* Phospholipids were determined as described by Rossi et al. (27).

TABLE III

Phospholipid Composition of Mammalian and Blowfly Flight Muscle Mitochondria

	% of total phospholipid			
	Liver	Heart	Kidney	Blowfly Muscle
Phosphatidylcholine	43	41	40	10–15
Phosphatidylethanolamine	34	37	38	60–70
Cardiolipin	17	19	19	15
Phosphatidylinositol	3	3	3	10

The data on liver, kidney and heart are taken from Fleischer et al. (23), those on blowfly muscle from a study by Carafoli et al. (26).

TABLE IV

Comparative Properties of the Different Ca^{++} Binding Processes

	Requirement for respiration or ATP hydrolysis	Maximal level (nMoles/ mg prot.)	Affinity (K_m)	Inhibition by	Biological distribution
Energy-linked uptake	yes	100–150	$\sim 10^{-6}$M	DNP, La^{+++}	absent from yeast, blowfly muscle*
High-affinity binding	no	0.2–11	$\sim 10^{-6}$M	DNP, La^{+++}	absent from yeast, blowfly muscle
Low-affinity binding	no	14–16	$\sim 10^{-4}$M	butacaine	absent from blowfly muscle
K^+ induced uptake	no	depends on the K^+ store	$\sim 10^{-4}$M	DNP, La^{+++}	not known

* See text four pages before this one.

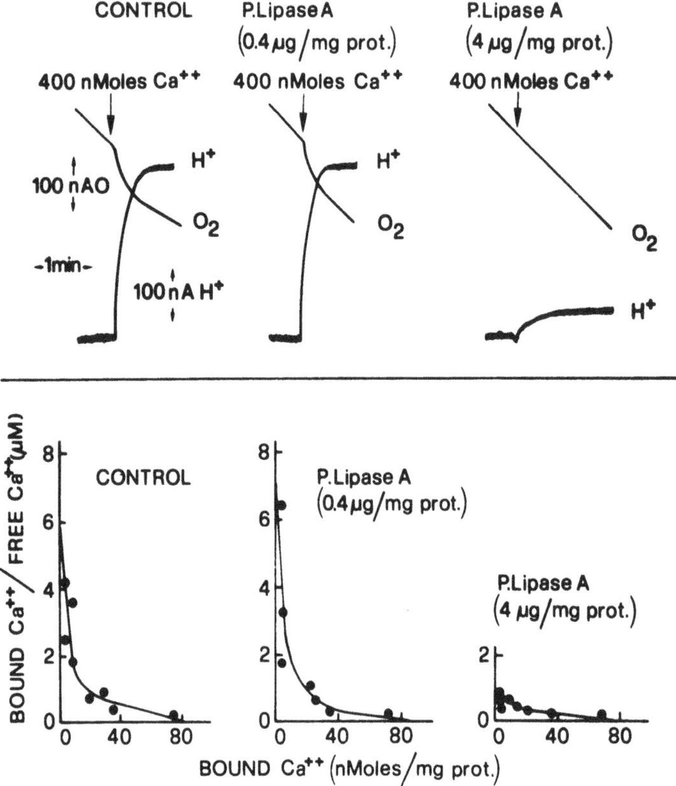

Fig. 11. Effect of phospholipase A on high-affinity Ca^{++} binding and energy-linked Ca^{++} uptake by mitochondria. Conditions as in Fig. 2,3 and 6. The incubation with phospholipase A was carried out at 21°, for 5 min., in 250 mM sucrose, with gentle stirring. At the end of the incubation, aliquots of the suspension containing the mitochondria were added to the reaction medium. Specific activity of phospholipase A:200 U/mg.

A summary of the main properties of the different Ca^{++} binding mechanisms is shown in Table IV. It is clear that energy-linked translocation, and high-affinity binding, share many properties. On the other hand, low-affinity binding, and energy-linked translocation, differ in many respects. What is then the role of the high-affinity binding, as reflecting the Ca^{++} carrier, and of the low affinity binding, as reflecting the surface binding of Ca^{++} to the mitochondrial phospholipids, in the energy-linked Ca^{++} translocation? The K^+-induced Ca^{++} uptake, undoubtedly very

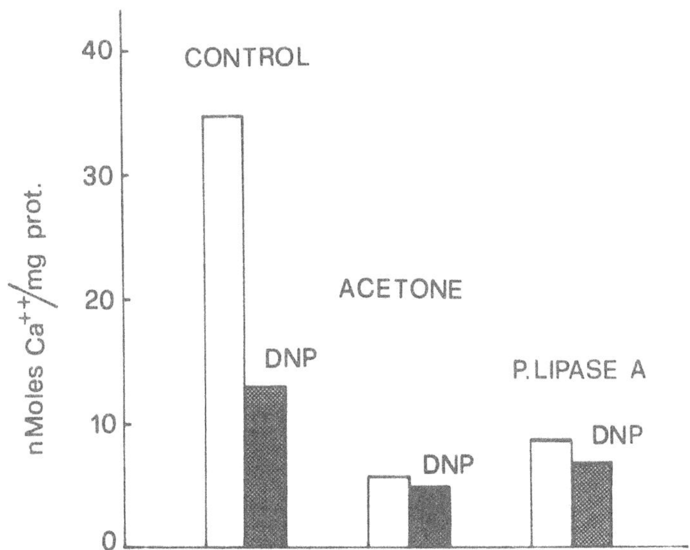

Fig. 12. Effect of phospholipase A and of acetone extraction on energy-linked Ca^{++} uptake and low-affinity Ca^{++} binding. Conditions as in Fig. 2. Concentration of DNP, 0.1 mM. Time of incubation, 1 min. The acetone treatment was carried out as described by Fleischer et al. (23). The treatment with phospholipase A was carried out at 0° for 20 min., with gentle stirring, in the following medium: 63 mM mannitol, 25 mM glycylglycine buffer pH 7.4, 0.13 % bovine serum albumin in glycylglycine buffer, 50 mM sucrose, and 200 mg mitochondrial protein. Volume, 20 ml. Phospholipase A, 0.1 mg. The incubation was terminated by the addition of 100 ml of 250 mM mannitol, 50 mM glycylglycine, 1 mM EDTA, 1% bovine serum albumin. Mitochondria were spun down and washed once with 250 mM mannitol.

interesting from a mechanistics point of view, is not considered in this context, since it requires the presence of valinomycin, a compound which is obviously foreign to normal mitochondria. The nature of the high-affinity compound must be considered first. A soluble protein component capable of high affinity Ca^{++} binding is presently under characterization in Lehninger's Laboratory (22) it thus seems very likely that the compound responsible for the high-affinity binding is a protein. However, its proper functioning apparently requires the integrity of the phospholipid structure

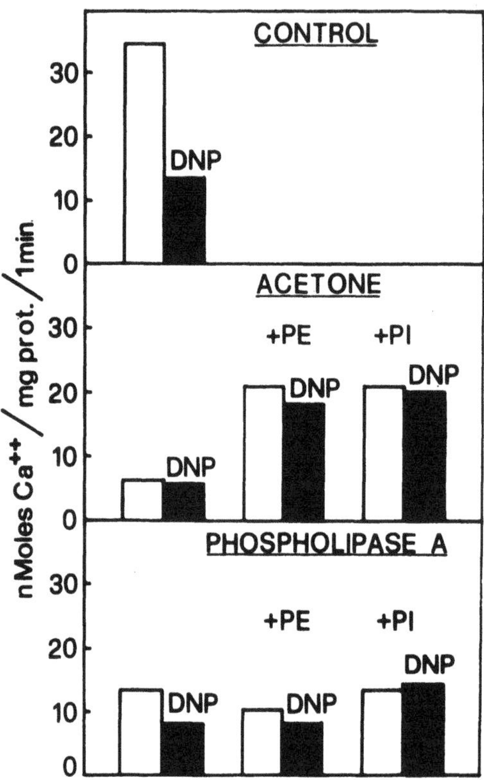

Fig. 13. Restoration of low-affinity binding in phospholipase-treated or acetone-extracted mitochondria by addition of phospholipids. Conditions as in Fig. 12. Mitochondria were treated with phosphatidylethanolamine or phosphatidylinositol as described by Scarpa and Azzone (6).

of mitochondria. Fig. 11 shows that a brief incubation of mitochondria with phospholipase A eliminates the high-affinity binding of Ca^{++}. It is interesting that the disappearance of the high-affinity reaction is paralleled by the disappearance of the energy-linked Ca^{++} translocation, as shown by the failure of Ca^{++} to induce the usual reversible shift of the respiratory rate and the ejection of protons. Under these mild conditions of incubation, the low-affinity binding is much less affected. When the degradation of the mitochondrial phospholipids is more severe, or when they are extracted with acetone (23), also the low-affinity binding decreases markedly, as shown in Fig. 12. Several attempts have

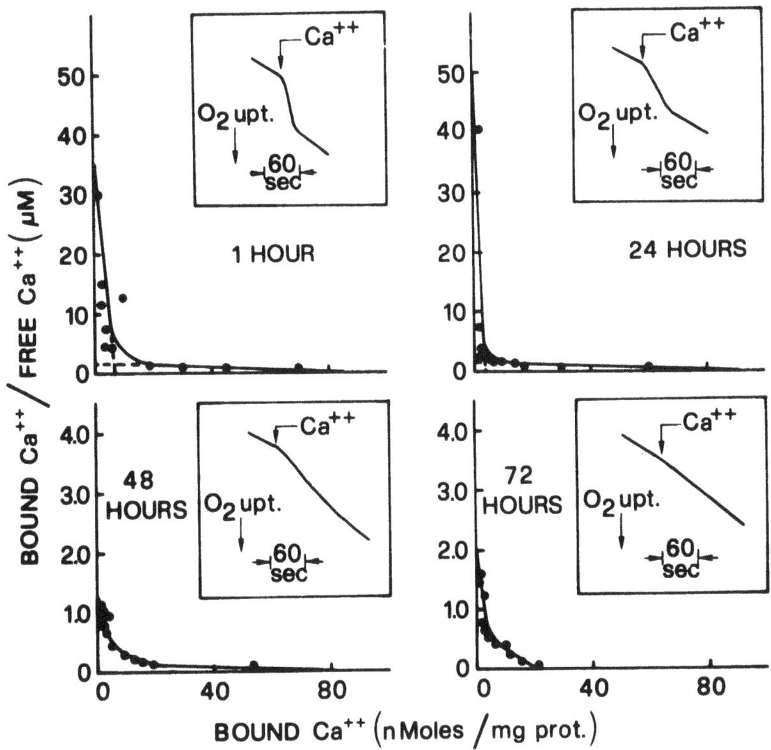

Fig. 14. Energy-linked Ca^{++} uptake, and high-affinity Ca^{++} binding in aged mitochondria. Conditions as in Fig. 2 and 6. Mitochondria were aged at 2°.

been made to restore the various Ca^{++}-binding processes by reincorporating phospholipids into mitochondria. As shown in Fig. 13, and as already demonstrated by Scarpa and Azzone (10), it was possible to restore low-affinity binding, but energy-linked translocation could never be restored. Also the attempts to restore high-affinity binding were completely unsuccessful. It is important to stress that commercial phospholipase A could affect also non-phospholipid components of the mitochondrial membrane. The conclusion that the integrity of the phospholipid structure is required for high-affinity Ca^{++} binding (and energy-linked Ca^{++} translocation) is thus at the moment still tentative.

As for the problem of the relationship among the different Ca^{++} binding mechanisms, there are several lines of evidence that

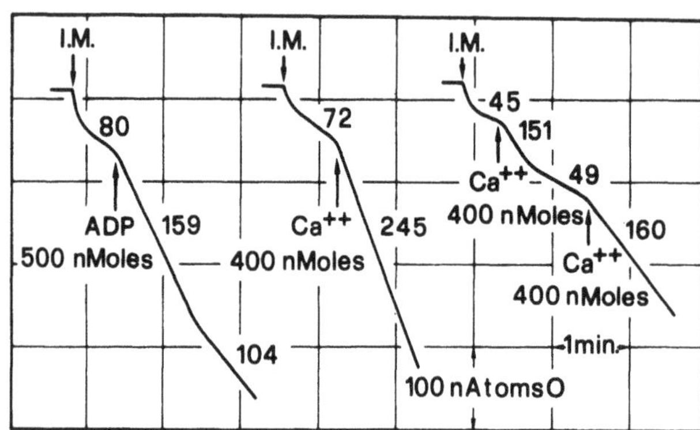

Fig. 15. Phosphorylation of ADP, and energy-linked Ca^{++} transport, in the inner membrane of mitochondria. Conditions as in Fig. 3 of Schnaitman and Greenawalt (24).

link the high-affinity reaction with the energy-linked translocation. The two systems have the same affinity for Ca^{++}, and are sensitive to the same inhibitor, La^{+++}. Uncouplers, which eliminate the energized state responsible for the active accumulation of Ca^{++}, also suppress the high-affinity reaction. As shown in Fig. 14, the disappearance of the energy-linked translocation in aged mitochondria is paralleled by the disappearance of the high affinity reaction. We have recently carried out a study of the distribution of the high and low-affinity binding sites, and of the energy-linked translocation of Ca^{++}, between the inner and outer mitochondrial membrane. The method of Schnaitman and Greenawalt (29) has been used, since it yields relatively intact preparations of the inner membrane, still containing most of the matrix; indeed, the respiration of our inner membranes was reversibly stimulated by Ca^{++}, both in the absence and in the presence of phosphate, as shown in Fig. 15. As expected from the literature (25), the stimulation was irreversible in the presence of phosphate. Thus, energy-linked Ca^{++} transport was present in the inner membrane, and it was on the other hand completely absent from the outer. The Scatchard plots shown in Fig. 16 show that both high and low-affinity binding are present in the inner membrane but only low-affinity binding is represented in the outer membrane. Thus, high-affinity reaction and energy-linked translocation have a parallel distribution, and it is clear on the other hand that the phospholipids of both membranes are involved in the low-affinity binding, despite the absence of energy-linked translocation from the outer membrane. Some other observations militate against a mandatory link between metabolism-independent, low-affinity binding and energy-linked translocation: for example, the already mentioned fact that the most important and effective inhibitor of the metabolism-independent binding, butacaine, does not inhibit the energy-linked uptake of Ca^{++}, rather it accelerates.

In conclusion, the bulk of the available evidence supports the role of the high-affinity reaction, and thus of the carrier, in the energy-linked translocation. Many interesting questions remain open; among them, of importance are the chemical nature of the Ca^{++} carrier, its molecular mechanism of action, and its functional link with the other ion transport systems in the mitochondrial membrane.

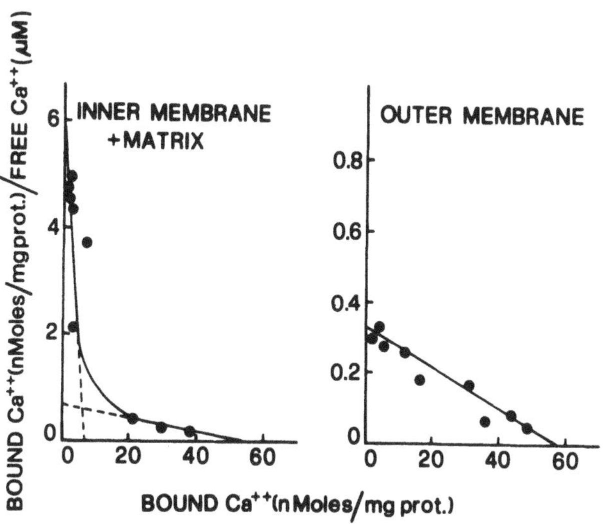

Fig. 16. High and low-affinity Ca^{++} binding in the inner and outer membrane of mitochondria. Conditions as in Fig. 6.

ACKNOWLEDGEMENTS

The Authors express their gratitude to Mrs. Daniela Ceccarelli-Stanzani for her skilled technical assistance in some of the experiments described.

REFERENCES

1. Lehninger, A.L., Carafoli, E., and Rossi, C.S., in "Adv. in Enzymol." (F.F. Nord, ed.), Vol. 29, p. 259, Interscience, New York (1967).
2. Rossi, C., Azzi, A. and Azzone, G.F., J.Biol.Chem. 242: 951 (1967).
3. Scarpa, A. and Azzone, G.F., Europ. J. Biochem. 12: 328 (1970).
4. Reynafarje, B. and Lehninger, A.L., J. Biol. Chem. 244: 584 (1969).
5. Greenawalt, J.W., Rossi, C.S. and Lehninger, A.L., J. Cell Biol. 23: 21 (1964).
6. Chance, B., J.Biol. Chem. 240: 2729 (1965).
7. Carafoli, E. and Azzi, A., unpublished observations.
8. Carafoli, E., J. Gen. Physiol. 50: 1899 (1967).
9. Scarpa, A. and Azzi, A., Biochim. Biophys. Acta, 150: 473 (1968).
10. Scarpa, A. and Azzone, G.F., Biochim. Biophys. Acta 173: 78 (1969).
11. Chance, B., Azzi, A. and Mela, L., in "The Molecular Basis of Membrane Function" (G. Tosteson, ed.), p. 561, Prentice-Hall, London (1969).
12. Mela, L., Arch. Biochem. Biophys. 123: 286 (1968).
13. Mela, L. and Chance, B., Biochem. Biophys. Res. Comm. 35: 556 (1969).
14. Carafoli, E., Gamble, R.L. and Lehninger, A.L., J. Biol. Chem. 241: 2644 (1966).
15. Lehninger, A.L., personal communication.
16. Fonyo, A. and Bessman, S.P., Biochem. Med. 2: 145 (1968).
17. Chappell, J.B., Brit. Med. Bull. 24: 150 (1968).
18. Lehninger, A.L. and Carafoli, E., in "Biochemistry of the Phagocytic Process" (J. Schulz, ed.), p. 9, North-Holland, Amsterdam (1969).
19. Carafoli, E. and Rossi, C.S., 1st International Symposium on

Cell Biology and Cytopharmacology, Venice, 1969, Raven Press, New York, in press.
20. Carafoli, E., Biochem.J. 116: 2P (1969).
21. Carafoli, E., Balcavage, W.X., Lehninger, A.L. and Mattoon, J. R., Biochim. Biophys. Acta, 205: 18 (1970).
22. Lehninger, A.L., personal communication.
23. Fleischer, S., Fleischer, B. and Stockenius, W., J. Cell Biol. 32: 193 (1967).
24. Schnaitman, C., Erwin, V.G. and Greenwalt, J.W., J. Cell Biol. 32: 719 (1967).
25. Rossi, C.S. and Lehninger, A.L., J. Biol. Chem. 239: 3971 (1964).
26. Carafoli, E., Hansford, G., Sacktor, B. and Lehninger, A.L., in press.
27. Rossi, C.R., Sartorelli, L., Tatò, L., Baretta, L. and Siliprandi, N., Arch. Biochem. Biophys. 118: 210 (1967).

PHOSPHOLIPASE A IN MITOCHONDRIAL MEMBRANES

P.M. Vignais, J. Nachbaur, A. Colbeau, P.V. Vignais

Laboratoire de Biochimie, Centre d'Etudes Nucléaires

et Faculté de Médecine, 38-Grenoble, France

The study of the distribution of phospholipase A in subcellular organelles represents a way of approach to the understanding of membrane permeability. As a typical example it may be mentioned that, under appropriate conditions, lysophosphatides which result from the attack of phospholipids by phospholipase A induce a transition of the phospholipid structure from a lamellar to a micellar organization (1,2). This micellar reorganization of a lipid structure favors the fusion of juxtaposed membranes as recently shown by Howell and Lucy (3) in the case of erythrocyte membranes. It may be speculated that such re-organization also occurs at the level of the tight junctions in plasma membranes which, according to Benedetti and Emmelot (4), display a typical hexagonal mosaic pattern, and is responsible for the high permeability of those junction areas (5).

The control by phospholipase A of membrane permeability can also be assessed in terms of a renewal of the fatty acid composition of membrane phospholipids occuring through the sequential action of phospholipase A and acyl-CoA transferase. It is known that the conformation of a phospholipid molecule depends on the nature of its fatty acid residues ; for instance, the mean area occupied per molecule in a monomolecular layer of phospholipids increases with the shortening of the hydrocarbon tails of the fatty acids or with their degree of unsaturation (6,7). That such a molecular alteration may influence the permeability of a membrane is suggested by the fact that artificial liposomes (8,9) become more permeable to glucose and glycerol when the degree of unsaturation of their phospholipids is increased.

Although for these reasons phospholipase A appears as a key enzyme in membrane permeability, it is evident that a lucid evaluation of its function depends on the correct assignment of its subcellular distribution. Unfortunately, this approach suffers from severe limitations due to still not completely perfect isolation of intact and pure subcellular organelles. In this communication, taking as examples mitochondria and mitochondrial membranes, we shall deal with the following problems :

1. What is the limitation of the currently available techniques for the clean separation of subcellular organelles and the isolation of pure membranes from these organelles ?

2. Is it possible to ascribe a phospholipase A activity to mitochondria by only correlating phospholipase A activity with the activity of typical marker enzymes of subcellular organelles ?

3. If phospholipase A is present in mitochondria, leading to a deacylation of mitochondrial phospholipids, it is highly probable that the counterpart mechanism, that is the lysophosphatide reacylation, is also accomplished by mitochondria. This feature will be briefly discussed.

I - COMMENTS ON THE SEPARATION AND PURIFICATION OF MEMBRANES FROM SUBCELLULAR ORGANELLES

Since phospholipase A as well as lysophospholipid reacylase activities have been detected in mitochondria, microsomes and lysosomes, we shall focus our attention on the separation of "pure" membranes from these particles.

Lysosomes may be obtained by the method described by Wattiaux (10) based on the fact that lysosomes are made lighter by <u>in vivo</u> incorporation of compounds of low density, for instance the non-detergent triton WR-1339. The purification of the triton-filled lysosomes (tritosomes) by centrifugation on a discontinuous sucrose gradient in view of phospholipase A assay has already been described (11-13). We only wish to draw attention to the fact that tritosomes can be disrupted by homogenization in an hypotonic medium, and that after centrifugation on a discontinuous sucrose gradient, their membrane collects at the same level as outer mitochondrial membrane (see below).

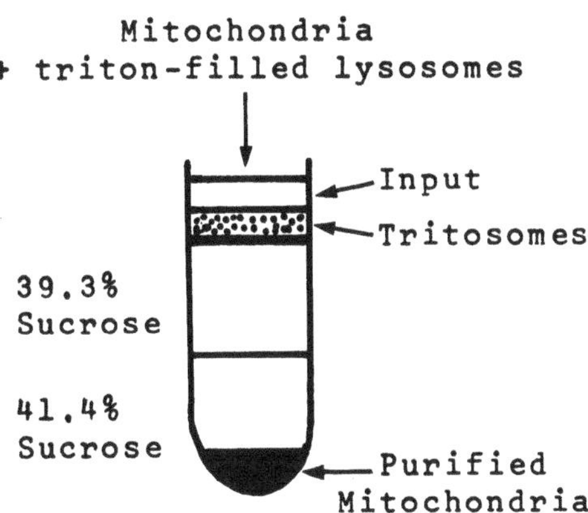

Fig.1 - Separation on sucrose gradient of mitochondria and triton-filled lysosomes.
The crude mitochondria preparation obtained by centrifugation at 5.000 g x 10 min. formed the input to the sucrose gradient and was spun at 23.000 r.p.m. for 3 hours. Particles floating on top of the gradient are triton-filled lysosomes (tritosomes). Mitochondria are found at the bottom of the tube.

Conversely, crude mitochondria preparations can be purified from contaminant triton-filled lysosomes by a preliminary centrifugation on sucrose gradient similar to that used to prepare the tritosomes.

The inner and outer mitochondrial membranes were isolated according to Parsons et al. (14) after swelling of mitochondria in 20 mM phosphate buffer. By this technique some outer membrane is found in the inner membrane fraction but on the other hand the contamination of the outer membrane fraction by inner membrane is very small. The inner membrane + matrix particles are disrupted by homogenization in a large volume of water or after sonication, and the inner membrane fraction can be easily separated on a discontinuous sucrose gradient (see Fig.3).

Microsomal membranes were isolated by the method of Molnar (15) on discontinuous sucrose gradient, in the presence of Mg^{++} for the rough endoplasmic reticulum membranes and the presence of EDTA for the smooth.

Table 1

Activities of marker enzymes in subcellular membranes

Fractions	Cyt.Ox. (o)	MAO (o)	Acid Ptase (+)	G-6-Ptase (+)	NADPH cyt. c red. (o)	5'AMPase (+)
Crude Mitoch.	1,120	6	57	37	3	61
Purified Mitoch.	1,830	6	21	30	6	22
Inner Memb.+ Ma	2,880	4	8	16	4	9
Outer Memb.	140	31	77	102	32	65
Tritosomes	10	<0.1	934	90	3	405
Tritos. Memb.	-	-	3900	690	-	-
Microsomes	<10	<0.1	37	190	82	61
Smooth ER	-	-	76	220	183	167
Rough ER	-	-	32	335	79	13

(o) nanomoles substrate oxidized/min/mg protein at 25°

(+) nanomoles P_i released/min/mg protein at 37°
(mean of two experiments).

Legend of Table 1 - Cytochrome oxidase was assayed spectrophotometrically at 25° andpH 7.4 (16). Since the oxidation of reduced cytochrome c is of first order with respect to its concentration, the rate constant k was deduced from a logarithmic plot of the kinetics of the reaction. The rate of oxidation of reduced cytochrome c (41µM) was calculated from the rate constant. Monoamine oxidase (MAO) was measured at pH 7.5 and at 25° by following the formation of benzaldehyde from benzylamine (17). Acid phosphatase was estimated with p-nitrophenyl-phosphate as substrate (18) at 37° after unmasking with Triton X-100. Glucose-6-Phosphatase was determined by the amount of inorganic phosphate formed at 37° and at pH 6.5 (19). NADPH-cytochrome c reductase was measured at 25° (20) and pH 7.8. 5'-nucleotidase (5'AMPase) was assayed according to Emmelot et al.(21) in the presence of Triton X-100 and at pH 7.5.
N.B. Except for glucose-6-phosphatase which is inhibited by Triton X-100 all the other enzymes were tested after lysis of membranes by Triton-X-100.

In Table 1 are given the mean average values of activities of some typical marker enzymes in "purified" mitochondria preparations, in microsomes, in tritosomes and in membranes derived from these particles. It is noteworthy that the glucose-6-phosphatase activity is predominant in the rough endoplasmic reticulum in agreement with Decloître and Chauveau (22) and NADPH cytochrome c reductase in the smooth endoplasmic reticulum in agreement with Manganiello and Phillips (23). The relatively high activity of glucose-6-phosphatase in mitochondria and in the outer mitochondrial membrane can be due in part to some unspecific breakdown of glucose-6-phosphate by acid phosphatase (24) bound to contaminant lysosomal membranes.

Brunner and Bygrave (25) have recently suggested that cytochrome P-450 would be a more suitable marker for microsomes than NADPH-cytochrome c reductase or glucose-6-phosphatase. Fig.2 shows that cytochrome P-450 is mainly localized in microsomes and especially in the smooth reticulum. However, the contamination of mitochondrial membranes, and especially of the outer membrane, by smooth endoplasmic reticulum is difficult to quantify by evaluation of the content in cytochrome P-450 (difference in absorbancy between 450 and 490 nm (26)). Only an upper limit of contamination of 20% of the outer mitochondrial membranes by microsomal membranes can be deduced from the spectra, a result which is in agreement with data obtained with the NADPH cytochrome c reductase.

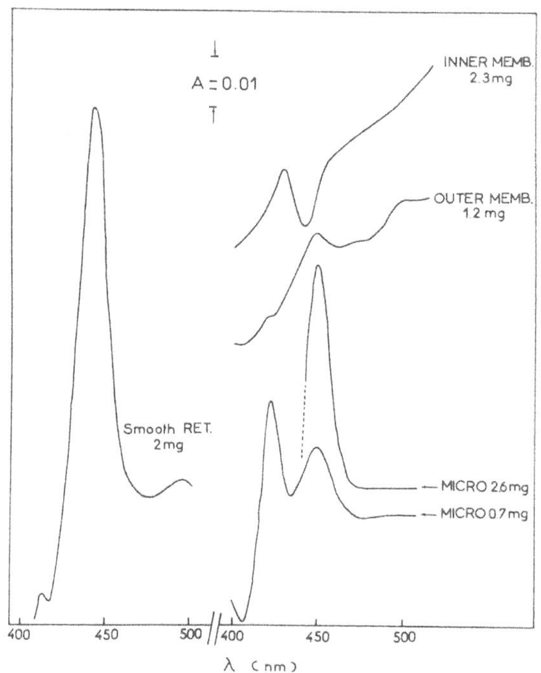

Fig.2 - Difference spectra of P-450 in subcellular membranes.

$Na_2S_2O_4$ (a few mg) was added to both cuvettes (containing 3 ml of the membrane suspension in 20 mM phosphate pH 7.5) and the content of one cuvette was bubbled with CO (O_2 free).

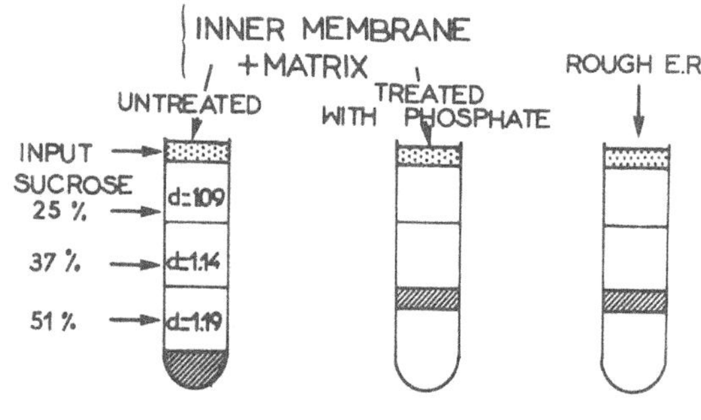

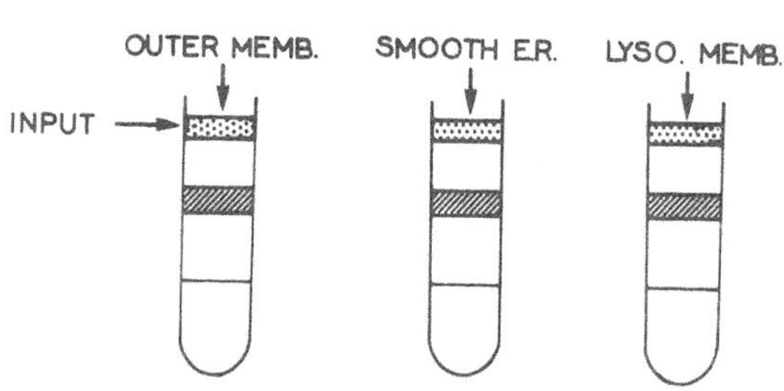

Fig.3 - Location of membranes from different types of subcellular organelles after centrifugation on 3 layer sucrose gradient (SW-39 ; 38.000 r.p.m.x45 min).

We shall consider now the behaviour of subcellular membranes centrifuged in the three-layer sucrose gradient that was used by Parsons (14) to separate the inner and the outer mitochondrial membranes. Such a discontinuous gradient is useful for an easy recovery of membranes at the level of the interfaces. However, it is clear (Fig.3) that its resolution power is limited, since the inner membrane (without matrix) sediments at the same level as the rough endoplasmic reticulum (second interface), and since the outer mitochondrial membrane, the smooth endoplasmic reticulum and the lysosomal membranes all gather at the same interface (between the two lighter sucrose layers). Indeed the buoyant density values of the membranes of the subcellular organelles fall in two groups : a first group comprising the rough endoplasmic reticulum (d = 1.19) and the inner mitochondrial membrane (d = 1.18) (which appears in the electron microscope as practically devoid of matrix but which still retain some malate dehydrogenase activity), the second comprising the outer mitochondrial membrane (d = 1.14), the smooth reticulum (d = 1.12) and the lysosomal membrane (d = 1.14) The buoyant densities values of these membranes are likely related to their phospholipid content (Table 2).

Table 2

Phospholipid content of subcellular fractions (μmole of lipid P/mg prot.)

Mitochondria	*Inner Memb.*	*Outer Memb.*	*Lysos. Memb.*	*Rough ER*	*Smooth ER*
0.16	*0.34*	*0.46*	*0.44*	*0.34*	*0.60*

After extraction of the lipids by the method of Dawson et al.(27) and solubilization by sulfuric acid the phosphorus content was determined by the method of Bartlett (28). Protein was determined by the biuret method.

We have also used yeast cells (Torulopsis utilis) for the study of the distribution of phospholipase A. After grinding the cells in a mannitol solution, according to Mattoon (29), the yeast homogenate was centrifuged at 600 g for 10 min.. The supernatant fluid was fractionated into four fractions by centrifugation at different speeds for different periods of time. Fraction I made of crude mitochondria was sedimented at 5,000 g for 15 min.. Fraction II was obtained by centrifugation at 20,000 g for 15 min. and Fraction III at 100,000 g for 1 hour. Fraction IV refers to soluble material not sedimenting at 100.000 g.

Cyt.oxidase(not corrected here for peroxidase) is found in Fractions I and II (Fig.4), glucose-6-phosphatase in the soluble fraction and also in Fractions II and III, and acid phosphatase and NADPH cytochrome c reductase are essentially in Fraction III. The high specific activity of NADPH cytochrome c reductase and acid phosphatase in Fraction III points to a localization of these two enzymes in a distinct class of subcellular organelles.

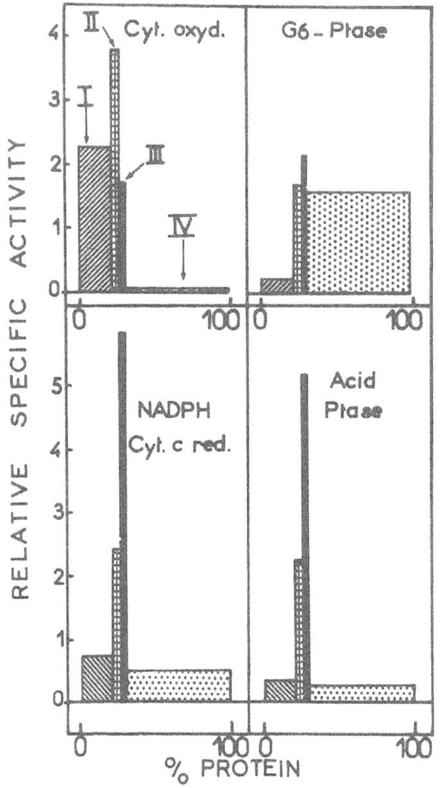

Fig.4 - Distribution pattern of enzymes in yeast cell (T.utilis). Fractions obtained by differential centrifugation of a yeast homogenate in 0.6 M mannitol are plotted in order of their isolation. The relative specific activity of each enzyme is referred to the specific activity found in the homogenate. On the abscissa each fraction is represented by its percentage of the total protein in the homogenate. Conditions of enzyme assays are as described in Table 1.

Crude mitochondria of Fraction I were washed 3 times with 0.6 M mannitol and further purified by centrifugation through a sorbitol gradient as described by South and Mahler (30) (Fig. 5). The mitochondrial particles gather in a well-limited layer which is enriched in cytochrome oxidase ; however, the specific activity of the mitochondria-bound glucose-6-phosphatase remains unchanged indicating that some glucose-6-phosphatase is strongly bound to mitochondria.

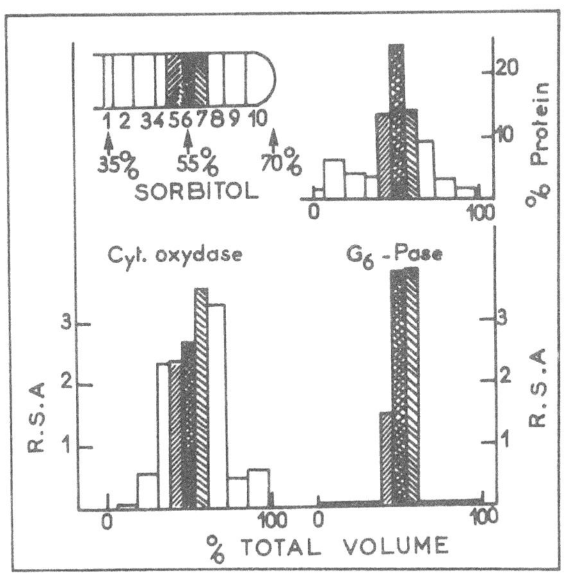

Fig. 5 - Isopycnic centrifugation of crude yeast mitochondria in a sorbitol gradient (30) (SW-39, 31,000 r.p.m. x 90 min.). On the abscissa each fraction is represented by its percentage of the total volume of sorbitol. The relative specific activity of each enzyme is referred to the specific activity of the starting crude mitochondria.

II - PHOSPHOLIPASE A IN MITOCHONDRIA

The chemical and biochemical aspects of the de-acylation of phospholipids by phospholipase A_1 and A_2 have been recently reviewed (31, 32). Beside the classical phospholipase A_2 which acts on the fatty ester linkage at the C_2-position of 3-phosphoglycerides, a phospholipase A_1, first identified in spleen by Lloveras et al. (33), attacks the ester in the 1-position of phosphoglycerides.

Summarized conclusions, taken from different papers, concerning the subcellular distribution of phospholipase A in mammalian cells are given in Table 3. Schematically, mitochondria contain a phospholipase A_2 (11, 13, 34-41), microsomes a phospholipase A_1 (37, 42, 43) which may be merely lipase (44), lysosomes a set of phospholipase A_1 and A_2 (11, 13, 38-40, 45-53) but no lysophospholipase (39). The brain phospholipase A_1 might be of lysosomal origin (53). Lysophospholipase is located in the cytosol of rat liver homogenates (36) (54), in the rough endoplasmic reticulum of bovine adrenal medulla (55), and in brain microsomes (56). Phospholipase A is not confined to mammalian cells, but is widespread in nature (see for review 31 and 32).

Table 3

Phospholipase distribution in subcellular organelles

MITOCHONDRIA (11,13,34-41)	*MICROSOMES (37,42,43,55, 56)*	*LYSOSOMES (11,13,38-40, 45-53)*	*CYTOSOL (36) (54)*
Plipase A_2	*Plipase A1*	*Plipase A_1*	*LysoPlipase*
	+ LysoPlipase	*Plipase A_2*	
Latency	*no Latency*	*A_1:pH opt.4-5 inhibited by Ca^{++}*	
pH opt. >8	*pH opt. > 8*		
Activated by Ca^{++}	*Insensitive to Ca^{++}*	*A_2:pH opt.6-8 inhibited by EDTA*	
Inhibited by ATP and Mg^{++}			
Located in the outer membrane			

We shall limit our discussion here to the mitochondrial phospholipase A, and we shall especially focus our attention on its discrimination from the lysosomal phospholipase A.

Between 1965 and 1967 several reports (34-37) mentioned the presence of a phospholipase A_2 in mitochondria and described the properties of that enzyme. The mitochondrial phospholipase A is activated by Ca^{++} and inhibited by EDTA ; its activity is optimum at alkaline pH and results in the removal of the fatty acids at the C_2-position. The latency period in the enzyme action may be due to the masking of phospholipase A in mitochondria since it can be shortened by freezing and thawing or by mitochondrial swelling induced by the addition of unsaturated fatty acids. Although these data were well documented, they were obtained with mitochondrial particles prepared according to the classical differential centrifugation technique without further purification.

Lysosomes also are endowed with a phospholipase A_2 activity (beside phospholipase A_1) and it was only 2 years ago that a series of papers by van Deenen and coworkers (40) and by ourselves (11, 13, 38) drew attention to the necessary discrimination between mitochondrial and lysosomal phospholipase A_2, a problem which has been revived by the recent claim (50) that lysosomal membranes display a high phospholipase A_2 activity at neutral pH.

Starting from rat liver mitochondria substantially free of lysosomes (eliminated as tritosomes, cf above) we have isolated mitochondrial membranes and we have compared the kinetics of accumulation at alkaline pH of lysophosphatidylethanolamine when either outer mitochondrial membrane or tritosomes are incubated with ^{32}P-labelled phosphatidylethanolamine. In an experiment summarized on Fig.6 and Table 4, the rate of accumulation of (^{32}P)-lysophosphatidylethanolamine at pH 8.5 was higher with the outer mitochondrial membrane than with tritosomes; in another experiment at pH 8.1, the rate of release of free fatty acids was about the same with the outer mitochondrial membrane as with lysosomal membrane (Table 5). These results indicate that, <u>at alkaline pH</u>, the specific activity of the phospholipase A bound to the outer mitochondrial membrane is, at least, equal to that found in lysosomes or in the lysosomal membrane.

Table 4

Kinetics of accumulation of (^{32}P)-lysophosphatidyl-ethanolamine

Fraction	Time min.	Tube n°	(^{32}P)-lysoPE formed (nmoles/mg prot.)
Mitochondria	Zero	Zero	6
	30	1	26
	60	2	38
	90	3	44
	120	4	43
Outer Mit. Memb.	Zero	20	6
	30	21	50
	60	22	64
	90	23	64
Tritosomes	Zero	30	5
	30	31	12
	60	32	17
	90	33	29

Conditions :

The substrate, (^{32}P)-phosphatidylethanolamine ((^{32}P)-PE), had been extracted from rat liver ^{32}P-labelled mitochondria.

0.3µmole of (^{32}P)-PE (24.000 c.p.m./µmole) emulsified by sonication was incubated at 37° with 2 mg of tissue protein in 0.05 M TRA buffer, pH 8.5, and 0,002 M $CaCl_2$ in a final volume of 1 ml.

The lipids were extracted by chloroform/methanol (2:1) and separated by thin-layer chromatography on silicic acid using, successively, first the chloroform/light petroleum/acetic acid (65:35:2 v/v) system, and then the chloroform/methanol/water (65:35:4 v/v) system (37). The autoradiography of the chromatoplate is presented in Fig. 6.

Fig.6-Radioautogram of the chromatoplate showing the accumulation of (^{32}P)-lysophosphatidylethanolamine with time.(cf Table 4).

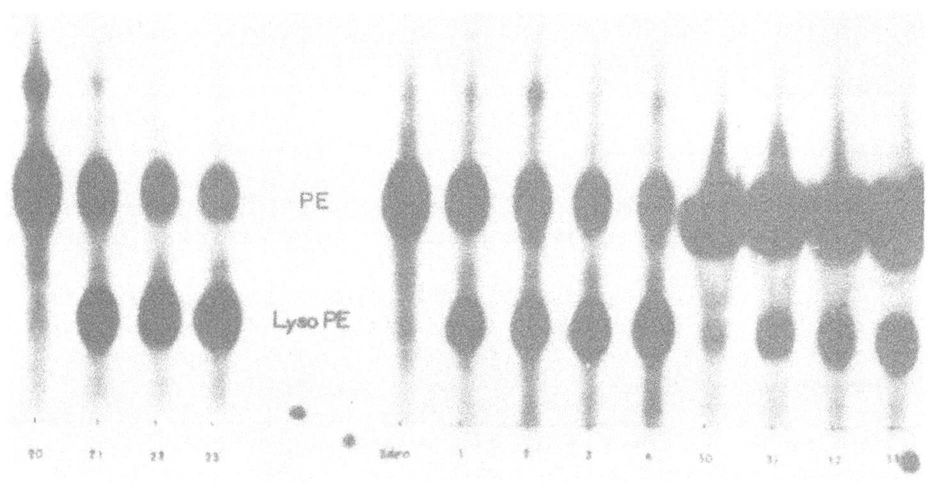

Table 5

Effect of pH on the phospholipase A activity of mitochondrial and lysosomal membranes

Fraction	pH	Fatty acids released: nmoles/hr/mg prot.	Fatty acids released: activity at pH 4.5 / activity at pH 8.1
Inner Memb.+ Ma	4.5	3	0.1
	8.1	25	
Outer Memb.	4.5	26	0.2
	8.1	110	
Lysosomal Memb.	4.5	266	2.2
	8.1	118	

(Reproduced from Biochem. Biophys. Res. Comm. (38)).

It has been shown by electron microscopy (13) that the soluble protein content of triton-filled lysosomes is partially replaced by triton ; on this basis, it is concluded that tritosomes are enriched in membrane components. Based on the specific activity of acid phosphatase which is an enzyme bound to the lysosomal membrane, the contamination of the outer mitochondrial membrane preparations by lysosomal membrane would amount to less than 10 per cent. Unless the lysosomal phospholipase is inhibited by the Triton WR-1339 still present in the lysosomes, it is not possible with these data to assign to the contaminant lysosomes the high phospholipase A_2 activity displayed by the outer mitochondrial membrane.

In the following experiments, we shall compare the substrate - and position - specificity and the optimum pH of phospholipase A in mitochondrial and lysosomal membranes.
As shown earlier (11, 13, 38) when purified mitochondria are incubated at pH 8.5 in the presence of Ca^{++}, endogenous phosphatidylethanolamine and phosphatidylcholine are virtually the only phospholipids to be hydrolyzed, phosphatidylethanolamine being the best substrate. The amount of lysoderivatives accumulated is practically equal to the amount of fatty acids released (13). Under these conditions, mitochondrial cardiolipin is not significantly hydrolyzed (13). It must be recalled that lysosomes efficiently hydrolyze cardiolipin with a rapid release of fatty acids (13, 51).

The mitochondrial phospholipase A is more active at alkaline pH than at acidic pH whereas the reverse is true for the lysosomal enzyme (Table 5).

The attack of egg phosphatidylethanolamine by the phospholipase A of the outer mitochondrial membrane at alkaline pH results in a release of more than 80% of unsaturated fatty acids. This effect is typical of a phospholipase A_2 since the fatty acids at the C_2-position in egg phosphatidylethanolamine are mostly unsaturated. In contrast with the outer mitochondrial membrane, the lysosomal membrane attacks egg phosphatidylethanolamine with a release of about 50% of unsaturated fatty acids over a large range of pH values (Fig.7).

The above data allow us to discriminate between the lysosomal and the mitochondrial phospholipase A and to ascribe a phospholipase A_2 activity to the outer mitochondrial membrane.

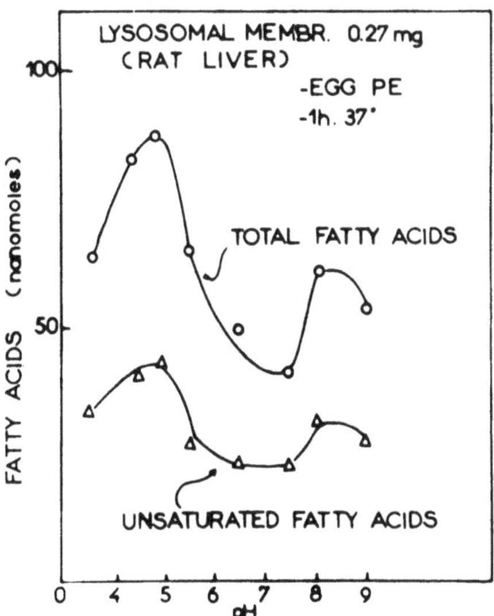

Fig.7 - Effect of pH on the release of fatty acids by lysosomal membranes. The tritosomes collected on top of the sucrose gradient (cf. Fig.1) were disrupted by homogenization in a large volume of 20 mM phosphate buffer pH 7.3. The lysosomal membranes were then purified by centrifugation on the 3-layer sucrose gradient shown on Fig.3. For the phospholipase test, 0.4 mg of egg phosphatidylethanolamine (egg PE) emulsified by sonication was incubated at 37° with 0.27 mg of lysosomal membrane protein in 0.002 M $CaCl_2$ and 0.05 M acetate buffer (pH 3.5 - 5.5), Tris, maleate buffer (pH 6.5 - 7.5) and Triethanolamine buffer (pH 7.5 - 9.0).

A phospholipase A activity has been described in yeast extracts (57) (58) and in yeast mitochondria (13). Phospholipase A activity of yeast mitochondria (Torulopsis utilis, Strain CBS 1516) has been tested with exogenous or endogenous phospholipids. When egg phosphatidylethanolamine is incubated in the presence of Ca^{++} with yeast mitochondria, lysophosphatidylethanolamine is accumulated mainly above pH 6 (Fig.8). At pH 4.5 where the accumulation of lysophosphatidylethanolamine is minimal, the release of free fatty acids is maximal, a result which means that the breakdown of phosphatidylethanolamine is pursued beyond the step of lysophosphatides.

It is known that the lipid material of mitochondria from yeast grown in semi-anaerobic conditions and in the

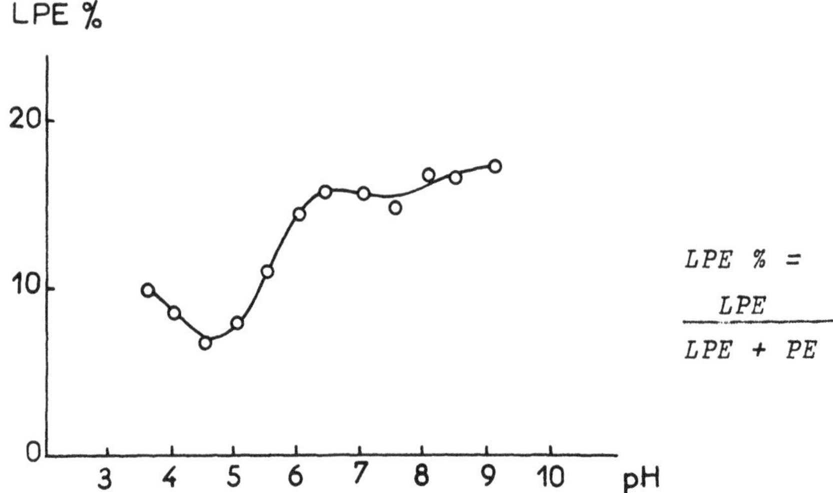

Fig.8 - Effect of pH on the phospholipase A of yeast mitochondria. 0.4 mg of egg PE emulsified by sonication was incubated with 0.5 mg protein of mitochondria and 2 mM $CaCl_2$ in 0.06 M acetate buffer (pH 3.8 to 5.5) or 0.06 M Tris, maleate buffer (pH 5.5 to 7.5) or 0.06 M Tris, HCl buffer (pH 7.5 to 8.8) with 0.5 mg protein of yeast mitochondria. Incubation for 30 min. at 37°.

presence of a high glucose concentration (10% w/v), is markedly altered (59, 60, 61). For instance, the ratio of the lipid phosphorus to the protein content drops from a value of 0.23 µmole of lipid P/mg of protein in mitochondria from aerobic cells to a value of 0.09 µmole in mitochondria from repressed cells (Table 6). The percentage of unsaturated fatty acids in mitochondrial phospholipids is also diminished in repressed cells ((59) and unpubl. results). In spite of these alterations in the lipid composition of mitochondria, there are no crucial differences in the phospholipase activity of normal and repressed yeast mitochondria.

Fig. 9 shows a kinetic study of the mitochondrial phospholipase A activity on endogenous phospholipids as estimated by the disappearance of both the total lipid phosphorus (left side) and the phosphatidylethanolamine (right side). The lipid phosphorus falls rapidly at pH 4.5 ; its remains constant at pH 8, although phosphatidylethanolamine is rapidly deacylated at this pH. This result is in agreement with the preferential accumulation of lysophosphatides at pH 8. The demonstration of the presence of a phospholipase A in yeast mitochondria and in mammalian mitochondria suggests that mitochondrial phospholipase A may be widely distributed among living species.

Table 6

Phospholipid composition of yeast mitochondria

	Cyt.Ox. (o)	Lipid P (+)	Per cent of lipid Phosphorus PC	PI	PE	CL
N-Mitoch.	3.0	0.23	42	9	35	12
R-Mitoch.	0.7	0.09	39	11	34	15

(o) μmole cyt.c oxidized/min/mg protein
(+) μmole lipid P/mg protein.

N-mitochondria were extracted from cells of ***T.utilis*** *(Strain CBS 1516) grown for 18 hr at 25° under forced aeration in fermenting jars containing 10 l of a medium made of 1% yeast extract, 2% peptone and 1.5% glucose (61).*

R-mitochondria were prepared from partially repressed cells grown in semi-anaerobic conditions and in the presence of a high concentration of glucose (10%).

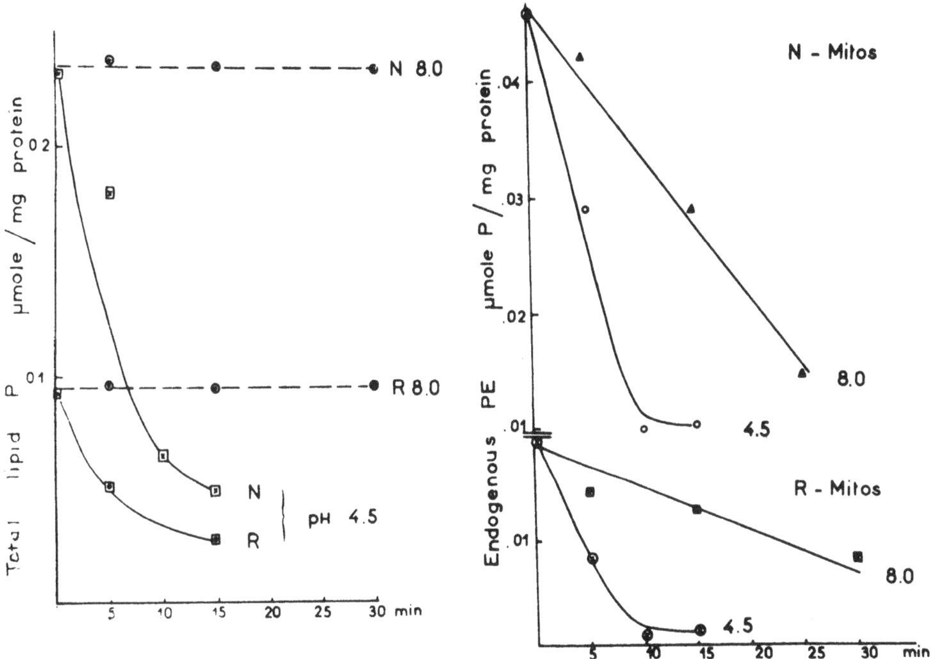

Fig. 9 - Activity of phospholipase A of yeast mitochondria on endogenous phospholipids. Yeast mitochondria (6.5 mg protein) were incubated at 37° in 0.06 M Tris, acetate buffer adjusted either at pH 4.5 or at pH 8.0 in a total volume of 2ml. After extraction of the total lipid content by chloroform/methanol (2:1) a phosphorus determination was made on an aliquote of the total lipid extract ; then PE was isolated by thin layer chromatography (cf Fig.6) and the spot analyzed for its phosphorus content.

III - POSSIBLE ENZYME ACTIVITIES ASSOCIATED TO PHOSPHOLIPASE A

If mitochondria are endowed with a phospholipase A it might be anticipated that they are also equiped with a reacylation system which allows, under physiological conditions, a reconstruction of their phospholipid material. When incubated with activated fatty acids, they might therefore incorporate the added fatty acids into their phospholipids. One may recall that the presence of a reacylating activity in the outer mitochondrial membrane has been reported by different laboratories (62-64). However, recently Eibl et al.(65) have found that the reacylating activity in liver cells is confined to microsomes.

We have reinvestigated this problem by measuring how the reacylating activity of a preparation of outer mitochondrial membrane is modified by additions of increasing amounts of microsomes. The relative amount of contaminant microsomes present in the outer mitochondrial membrane preparation was calculated according to its glucose-6-phosphatase activity. Extrapolation to a zero glucose-6-phosphatase activity (i.e. to a zero microsomal contamination) shows that no (^{14}C)-oleate incorporation is detectable in the phosphatidylcholine of outer mitochondrial membrane theoretically devoid of microsome contaminant (fig.10). Based on the same rationale the amount of (^{14}C)-oleic acid incorporated into phosphatidylethanolamine was found negligible.

In these experiments an oleoyl-CoA generating system made of ATP, CoA, (^{14}C)-oleic acid and outer mitochondrial membrane rather than (^{14}C)-oleoyl-CoA was used and one must keep in mind that although the outer mitochondrial membrane possesses the ability to activate long chain fatty acids (66), the rate of activation may be limiting in the overall process of lysophosphatide reacylation.

Another remark concerns the labelling of mitochondrial phospholipids, other than phosphatidylcholine and phosphatidylethanolamine, by (^{14}C)-oleic acid.
A significative incorporation of (^{14}C)-oleic acid, not accounted for by microsome contamination, was detected in two compounds having the same Rf as phosphatidic acid and phosphatidyl-glycerol on thin layer of silicic acid. The reacylation of lysophosphatidic acid in the outer mitochondrial membrane has already been reported by Zborowski and Wotjczak (67), by Shephard and Hübscher (68)

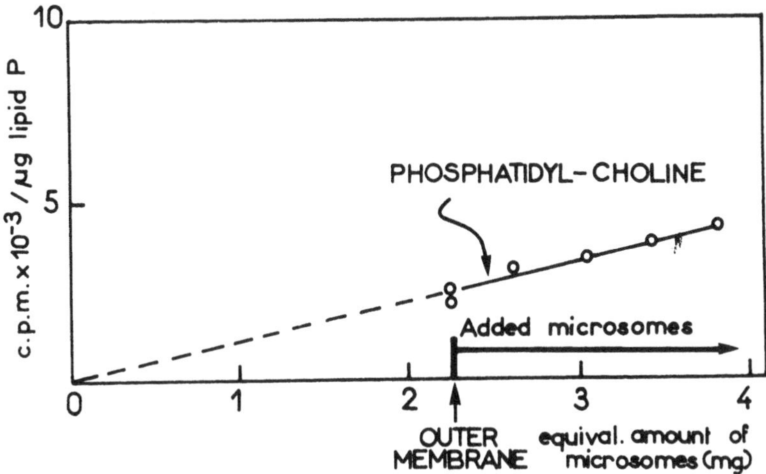

Fig.10 - Incorporation of (^{14}C)-oleic acid into the phosphatidylcholine of an outer mitochondrial membrane preparation supplemented with various amounts of microsomes. The medium was made of 0.2 μmole (^{14}C)-oleic acid (2,150 c.p.m./nanomole), 2 mM ATP, 12.5 mM $MgCl_2$, 2.5 mM $CaCl_2$, 0.4 mM CoA, 50 mM FNa, in 40 mM TRIS-HCl buffer pH 8. Incubation was carried out for 20 min. at 37° with 6 mg (protein) of the outer mitochondria membrane preparation alone or supplemented with various amounts of microsomes as indicated on the Fig.- Total volume 2 ml - The amount of contaminant microsomes in the outer mitochondrial membrane preparation (6 mg) was approximated to 2.3 mg as assessed by the glucose-6-phosphatase activity. The incubation was stopped by addition of 4 ml of a mixture of chloroform/methanol (2/1 - v/v) and phospholipids were isolated after chromatography on their layer of silicic acid.

and by Sarzala et al. (64).

Since the outer membrane of rat liver mitochondria do not seem to reacylate at an appreciable rate its endogenous lysophosphatidylethanolamine and lysophosphatidylcholine formed through phospholipase A activity, one may wonder by what mechanism these lysoderivatives are eliminated from the outer mitochondrial membrane. One possibility would be the replacement of the mitochondrial lysoderivatives by microsomal phospholipids by means of an exchange mechanism similar to that described in the case of mitochondrial and microsomal phospholipids (69, 70, 71).

Acknowledgments

The authors gratefully thank Mme J. Chabert, M. R. Césarini and Mme M. Bof for their excellent technical assistance.
This work was supported by grants from the "C.N.R.S." (E.R.A. n° 36), from the "Fondation pour la Recherche Médicale", and from the "D.G.R.S.T.".

REFERENCES

1. Bangham, A.D. and Horne, R. W., J. Mol. Biol. 8:660 (1964).
2. Haydon, D. A. and Taylor, J., Theoret. Biol. 4:281 (1963).
3. Howell, J. J. and Lucy, J. A., FEBS-Letters, 4:147 (1969).
4. Benedetti, E. L. and Emmelot, P., in "Ultrastructure in Biological Systems" (A.J. Dalton and F. Hagueneau, Eds.) Vol. 7, p.33, Academic Press, London and New-York (1968).
5. Loewenstein, W. R., Annals N. Y., Acad. Sci. 137:441 (1966).
6. Demel, R. A., Van Deenen, L. L. M. and Pethica, B. A., Biochim. Biophys. Acta, 135:11 (1967).
7. Van Deenen, L. L. M., in "The molecular basis of membrane function" (D.C. Tosteson, Ed.), p. 47, Prentice Hall, New Jersey (1969).
8. Bangham, A. D., Standish, M. M. and Watkins, J. C., J. Mol. Biol. 13:238 (1965).
9. Bangham, A. D., De Gier, J. and Greville, G. D., Chem. Phys. Lipids, 1:225 (1967).
10. Wattiaux, R., Wibo, M. and Baudhuin, P., in "Lysosomes" (A. V. S. De Reuck and M. P. Cameron, Eds.), p. 176, Ciba Foundation Symposium (1963).
11. Vignais, P. M. and Nachbaur, J., Bull. Soc. Chim. Biol. 50:1473 (1968).
12. Vignais, P. M. and Nachbaur, J., Biochem. Biophys. Res. Comm. 33:307 (1968).
13. Vignais, P. M., Nachbaur, J., André, J. and Vignais, P. V., in "Mitochondria, Structure and Function" (L. Ernster and Z. Drahota, Eds.) Vol. 17, p.43, FEBS-Symposium, Academic Press, London and New-York (1969).
14. Parsons, D. F., Williams, G. R., Thompson, W., Wilson, D. and Chance, B., in "Mitochondrial Structure and compartmentation" (E. Quagliariello, S. Papa, E. C. Slater and J. M. Tager, Eds.), p.9, Adriatica Editrice Bari (1967).

15. Molnar, J., Biochemistry, 6:3067 (1967).
16. Appelmans, F., Wattiaux, R. and de Duve, C., Biochem. J. 59:438 (1955).
17. Tabor, C. W., Tabor, H. and Rosenthal, S. H., in "Methods in Enzymology" (S. P. Colowick and N. O. Kaplan, Eds.), Vol. 2, p. 390, Academic Press, London and New-York (1955).
18. Linhardt, K. and Walter, K., in "Methods of Enzymatic Analysis" (H. U. Bergmeyer, Ed.) 2nd Edition, p.783, Verlag Chemie.(1965).
19. de Duve, C., Pressman, B. C., Gianetto, R., Wattiaux,R. and Appelmans, F., Biochem. J. 60:604 (1955).
20. Omura, T., Siekevitz, P. and Palade, G. E., J. Biol. Chem. 242:2389 (1967).
21. Emmelot, P., Bos, C. J., Benedetti, E. L. and Rümke, Ph., Biochim. Biophys. Acta, 90:126 (1964).
22. Decloître, F. and Chauveau, J., Bull. Soc. Chim. Biol. 50:491 (1968).
23. Manganiello, V. C. and Phillips, A. P., J. Biol. Chem. 240:3951 (1965).
24. Leighton, F., Poole, B., Beaufay, H., Baudhuin, P., Coffey, J. W., Fowler, S. and de Duve, C., J. Cell. Biol. 37:482 (1968).
25. Brunner, G. and Bygrave, F. L., European J. Biochem. 8:530 (1969).
26. Omura, T. and Sato, R., J. Biol. Chem. 239:2370 (1964).
27. Dawson, R. M. C., Hemington, N. and Lindsay, D. B., Biochem. J., 77:226 (1960).
28. Bartlett, G. R., J. Biol. Chem. 234:466 (1959).
29. Mattoon, J. R. and Sherman, F., J. Biol. Chem. 241:4330 (1966).
30. South, D. J. and Malher, H. R., Nature 218:1226 (1968).
31. Van Deenen, L. L. M. and de Haas, G. H., Ann. Rev. Biochem. 35:157 (1966).
32. Goldfine, H., Ann. Rev. Biochem. 37:303 (1968).
33. Lloveras, J., Douste Blazy, L. and Valdiguie, P., Compt. Rend. 256:1851 (1963).
34. Rossi, C. R., Sartorelli, L., Tato, L., Baretta, L. and Siliprandi, N., Biochim. Biophys. Acta, 98:207 (1965).
35. Scherphof, G. L. and Van Deenen, L. L. M., Biochim. Biophys. Acta, 98:204 (1965).
36. Bjørnstad, P., Lipid Res. 7:612 (1966).
37. Waite, M. and Van Deenen, L. L. M., Biochim. Biophys. Acta, 137:498 (1967).
38. Nachbaur, J. and Vignais, P. M., Biochim. Biophys. Res. Comm. 33: 315 (1968).

39. Stoffel, W. and Greten, H., Hoppe Seyler's physiol. chem. 348:1145 (1968).
40. Waite, M., Scherphof, G. L., Boshouwers, F. M. G. and Van Deenen, L. L. M., J. Lipid Res. 10:411 (1969).
41. Waite, M., Biochemistry, 8:2536 (1969).
42. Bjørnstad, P., Biochim. Biophys. Acta, 116:500 (1966).
43. Scherphof, G. L., Waite, M. and Van Deenen, L. L. M., Biochim. Biophys. Acta, 125:406 (1966).
44. Slotboom, A. J., de Haas, G. H., Bonsen, P. P. M., Burdach-Westerhuis, G. J. and Van Deenen, L. L. M., Chem. Phys. of Lipids, 4:15 (1970).
45. Mellors, A. and Tappel, A. L., J. Lipid Res. 8:479 (1967).
46. Mellors, A., Tappel, A. L., Sawant, P. L. and Desai, I. D., Biochim. Biophys. Acta, 143:299 (1967).
47. Smith, A. D. and Winkler, H., Biochem., J. 108:867 (1968).
48. Blaschko, H., Smith, A. D., Winkler, H., Van den Bosch, H. and Van Deenen, L. L. M., Biochem. J. 103: 30C (1967).
49. Winkler, H., Smith, A. D., Dubois, F. and Van den Bosch, H., Biochem. J. 105:38C (1967).
50. Rahman, Y. E., Verhagen, J. and Wiel, D., Biochim. Biophys. Res. Comm. 38:670 (1970).
51. Fowler, S. and de Duve, C., J. Biol. Chem. 244:471 (1969).
52. Lloveras, J. and Douste Blazy, L., Bull. Soc. Chim. Biol. 51:67 (1969).
53. Gatt, S., Biochim. Biophys. Acta 159:304 (1968).
54. Van den Bosch, H., Aarsman, A. J., Slatboom, A. J. and Van Deenen, L. L. M., Biochim. Biophys. Acta, 164:215 (1968).
55. Hörtnagl, H., Winkler, H. and Hörtnagl, H., Europ. J. Biochem. 10:243 (1969).
56. Leibovitz, Z. and Gatt, S., Biochim. Biophys. Acta, 164:439 (1968).
57. Van den Bosch, H., Van der Elzen, H. M., and Van Deenen, L. L. M., Lipids, 2:279 (1967).
58. Letters, R., in "Aspects of yeast metabolism" (A. K. Mills and A. H. Krebs, Eds.) p. 303, Blackwell Scientific Publications, Oxford (1967).
59. Jollow, D., Kellerman, G. M. and Linnane, A. W., J. Cell. Biol. 37:221 (1968).
60. Luckins, H. B., Jollow, D., Wallace, P. G. and Linnane, A. W., Aust. J. Exp. Biol. Med. Sci. 46:651 (1968).
61. Vignais, P. M., Nachbaur, J., Huet, J. and Vignais, P. V., Biochem. J. 116:42P (1970).
62. Stoffel, W. and Schiefer, H. G., Hoppe Seyler's Physiol. Chem. 349:1017 (1968).

63. Nachbaur, J., Colbeau, A. and Vignais, P. M., FEBS-Letters 3:121 (1969).
64. Sarzala, M. G., Van Golde, L. M. de Kruyff, B. and Van Deenen, L. L. M., Biochim. Biophys. Acta 202:106 (1970).
65. Eibl. H., Hill, E. E. and Lands, W. M. E., European J. Biochem. 9:250 (1969).
66. Norum, K. R., Farstad, M. and Bremer, J. Biochem. Biophys. Res. Comm., 24:797 (1966).
67. Zborowski, J. and Wojtczak, L., Biochim. Biophys. Acta 187:73 (1969).
68. Shephard, E. H. and Hübscher, G., Biochem. J. 113:429 (1969).
69. Wirtz, K. W. A. and Zilversmit, D. B., J. Biol. Chem. 238:2615 (1968).
70. Mc Murray, W. C. and Dawson, R. M. C., Biochem. J. 112:91 (1969).
71. Akiyama, M. and Sakagami, T., Biochim. Biophys. Acta, 187:105 (1969).

MEMBRANE-BOUND ENZYMIC ACTIVITY IN THE BASE-EXCHANGE REACTIONS OF PHOSPHOLIPID METABOLISM

G. PORCELLATI and F. di JESO

Istituti di Chimica Biologica, Università di Pavia e Perugia, Italy

INTRODUCTION

A number of observations has been recently made of calcium-stimulated base exchange reactions for phospholipid synthesis, mostly in non-nervous tissues (1-6). The enzymic reactions are believed to take place predominantly in the microsomal particles, and lead to the non-net synthesis of lipid at the expenses of pre-existing endogenous phospholipid molecules.

Few preliminary results have reported the possibility that such base-exchange reactions could take place in brain (7-9), but no detailed experimental evidence has been as yet produced in support of these reactons. We have recently obtained evidence (10, 11) for a calcium-dependent incorporation of L-3-^{3}H-serine and 1,2-^{14}C-ethanolamine into the phospholipid of isolated microsomal membranes from chick brain. A membrane-bound activity has been detected in noticeable amounts, which converts into lipid material about 120 nmoles of labelled serine or 200 nmoles of radioactive ethanolamine/g of fresh brain microsomes/hr, when assayed under optimal conditions.

The present report studies additional properties of the system obtained from the microsomal membranes of chick brain, examines possible enzymic mechanisms, and gives some insight as regard to the possibilities of solubilizing the enzymic system.

RESULTS AND DISCUSSION

General Properties of the Membrane-Bound System

The extent of serine or ethanolamine incorporation into microsomal lipid by calcium-dependent base-exchange reactions has been found to be proportional to the concentration of the microsomal suspension, and proceedes linearly with time up to 30 min of incubation. The rate of incorporation levells off after that time. It is not known from our experiments whether this kind of time-activity relationship is due to the attainment of equilibrium by the reaction(s).

In studies using radioactive ethanolamine one must distinguish between incorporation via the cytidine pathway (see 12,13) and the exchange between free ethanolamine and phospholipids. That the enzymic incorporation examined in this study is merely due to the latter mechanism was proved by the results of various experiments here described : (a), supplementing the standard incubation system, reported in Fig.1, with 0.5 to 2 mM phosphorylethanolamine (PE), either separately or together with 2 mM cytidine triphosphate (CTP), did not influence at all the enzymic activity, thus indicating that the transfer of ethanolamine was not "diluted out" by added unlabelled PE ; (b), substituting the radioactive ethanolamine with ^{14}C-labelled-PE (1.6 mM, Specific Activity of 0.2 microcurie/micromole) and CTP (4 mM) was ineffective ; (c), adding separately to the standard incubation mixture 0.5 to 3 mM ATP, 0.5 to 2 mM cytidine monophosphate (CMP), 1 to 2 mM <u>sn</u>-glycero-3-phosphoric acid (GPA), 1,2-diglyceride from ovolecithin (0.2 to 2 mM) and 0.2 to 1 mM 1,2-diacyl GPA (phosphatidic acid), either of which might have acted as acceptors for the possibly activated form of ethanolamine, was also ineffective. These two last lipids were added to the incubation system in the presence of minimal amounts of calcium ions, to prevent their precipitation. These results, together with the consideration that the presence of Ca^{++} ion in the enzymic system would have invariably produced a serious inhibition of the PE:diglyceride transferase (E.C. 2.7.8.2), should this enzyme be present in the subcellular fraction, point to the conclusion that the incorporation of ethanolamine reported in this section is only due to the calcium-mediated exchange system.

The rate of serine incorporation into chick brain phospholipids

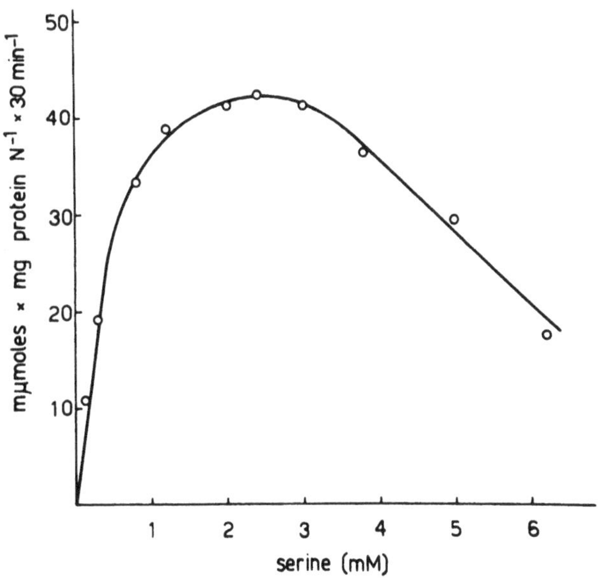

Fig.1 - The Effect of L-Serine Concentration on the Incorporation of Serine into the Microsomes of Chick Brain Tissue in vitro. Microsomal membranes from the brains of 4-9 days-old chicks were prepared and purified, as explained elsewhere (12-14). The pellet constituting the microsomes was washed once, in about 1/5 of the volume of the original homogenate, in a chilled 0.32 M-sucrose-2 mM-beta-mercaptoethanol solution, recentrifuged for 60 min at 105,000 x g_{av}, taken up in small volumes of the same solution, and analysed for the presence of some marker-enzymes and chemical constituents (8, 12-14).

The labelling of phospholipids was measured in a Tris-buffered medium, containing 40 mM-Tris-HCl buffer, pH 8.0, 4 mM-cysteine, L-3-^{3}H-serine (780 nc) added as shown, 0.9 mg of microsomal protein (about 100 mg of original fresh tissue), suspended in 0.32 M-sucrose-2 mM-beta-mercaptoethanol, and 25 mM-calcium chloride. Enzyme and other components were added at the indicated order. The final volume was 0.5 ml. The tubes were incubated for 30 min, at 37°. After incubation, the reaction mixture was inactivated by adding ice-cold 50 % trichloroacetic acid (TCA) to give a final concentration of 5 %. After washing with ice-cold, unlabelled L-serine (2 mM), protein-lipid linkages were disrupted, lipid was extracted from the precipitate and purified, following previous indications (15,16). Known portions of the radioactive lipid solution were dried directly into the scintillation vials, and counted (12-14). Activity is expressed as nmoles serine/mg protein N/30 min.

per g of original fresh tissue microsomes is rather high, the per cent of conversion being of the order of 10 % of the incubated serine, under optimal conditions. Similarly the rate of ethanolamine incorporation is of the order of about 5 % of the incubated base.

When the concentration of L-serine is varied over more than a 20-fold range, the enzymic system for the exchange of serine is apparently saturated by about 2.4 mM substrate (Fig.1). A noticeable decrease in velocity appears with increase of substrate concentration. This result may indicate either that there is a substrate inhibition of the enzyme or that the whole enzymic activity may be limited by the fact that the exchanging molecules of phospholipids, i.e. the microeomes, are kept constant in the experimental assay, whereas the amounts of serine are increasing. Similar findings have been obtained by preparing the microsomes in an EDTA-containing medium. However, by ignoring this effect at higher concentrations of serine, a K_m value of 3.3×10^{-4} M has been determined from the results of Fig.1, with a V_{max} of about 45 nmoles per mg of protein N per 30 min.

When the microsomal preparation from the chick brain was incubated with varying amounts of ethanolamine under similar conditions as described in Fig.1, a substrate concentration-velocity relationship was obtained, which is similar in many respects to that observed with L-serine. The system is apparently saturated by about 2.2 mM ethanolamine, with a noticeable decrease in velocity with increase in substrate concentration above that level. By ignoring this effect at higher concentrations of ethanolamine, a K_m value of about 3.0×10^{-4} M has been obtained. This value is not unlike that of 0.20 mM reported by CRONE (5).

The effect of pH on serine incorporation is reported in Fig.2. The optimum lies in a broad region between pH 8.0 and pH 9.0, with a slow fall-off in activity at pH values above 8.9-9.0. Large activity (80 % of the previously mentioned optimum activity) is still evident at pH 10. Similar results have been obtained by replacing Tris buffer with a veronal buffer. As regard to the ethanolamine exchange, an optimum pH value has been found around pH 8.5, in close agreement with the findings of VANDOR and RICHARDSON (6), who worked with plant microsomes. A much lower activity exists below pH 6.5 and above pH 10. Similar results have been obtained by replacing the Tris buffer with a Veronal buffer.

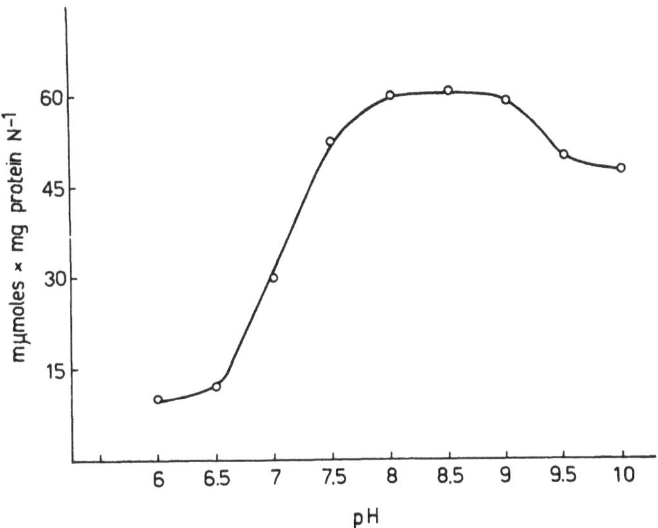

Fig.2 - The Effect of pH on Serine Incorporation into Phospholipid of Chick Brain Microsomes in vitro.
The reaction mixture is similar to that reported in Fig.1, except that 20 micromoles of the Tris-HCl buffer solution was adjusted to the indicated pH values with NaOH, and 2.2 mg of microsomal protein was used. The final concentration of substrate (780 nc) was 2 .4 mM, and the time of incubation of 30 min.
Rate of incorporation was determined, as reported in Fig.1, and is expressed as nmoles of serine incorporated/mg of protein N.
The pH of the incubation mixture was determined with a glass electrode after the addition of all reaction components.

Effect of Calcium and Magnesium Ions on the Rate of Exchange of Serine and Ethanolamine

It has been already mentioned before the requirement for Ca^{++} in the serine-incorporating system. The requirement is already detectable at a calcium concentration as low as 0.05 mM (Table I,a), and is more easily evidentiated, if the brain microsomes are prepared by homogenization in sucrose containing 10 mM-EDTA and sedimentation, followed by resuspension in an EDTA-free sucrose solution and re-sedimentation. A 12-fold increase of activity is thus observable by the addition of calcium to 1 mM. On the contrary, a proportional calcium-dependent increase is less evident, if the microsomes are initially prepared in an EDTA-free medium, as shown in

TABLE I.

The Effect of Adding Calcium Ions on the Incorporation of Labelled Serine into the Phospholipids of Brain Microsomes

Ca^{++} (mM)	Activity (a)	Activity (b)
-	4.4	0.9
0.05	5.3	1.0
0.1	6.5	2.5
0.5	9.1	8.9
1.0	10.3	12.1

In the experiments (a), the microsomes were prepared in 0.32 M-sucrose-2 mM-beta-mercaptoethanol solution, while in the experiments (b), they were prepared on a 0.32 M-sucrose-2 mM-beta-mercaptoethanol-10 mM-EDTA solution, resuspended in the same solution without EDTA, and again sedimented.
Activity is expressed as nmoles of labelled lipid x mg protein N^{-1} x 30 min^{-1}. Experimental conditions were similar to those reported in Figs.1 and 2.

Table I. This is apparently due to a small incorporating activity attributable to traces of endogenous calcium ions present in the microsomal preparations. Accordingly, by adding 2 mM-EDTA to the reaction mixture incubated with the microsomes prepared in an EDTA-free medium, we have observed that the residual activity is lowered to extremely low values, which are comparable to the levels obtained with the microsomes prepared in an EDTA-containing sucrose solution.

The maximum incorporation of serine is reached at a Ca^{++} ion concentration between 25 and 30 mM (Fig.3) in both Tris and Veronal buffers. At calcium values of about 1 mM, the incorporating activities are approximately similar, either for the microsomes prepared in an EDTA-containing medium or for those centrifuged without the presence of EDTA in the sucrose (see also Table I). If EDTA is added directly to the incubation mixture (5 mM), detectable activity is evident only when the calcium concentration is higher than 2 mM (Fig.3) ; the incorporation rates reach thereafter almost the standard levels, when calcium concentration is increased up to 10 mM. Similar results have been obtained if the brain microsomes are pre-incubated with 5 mM-EDTA for 15 min in standard conditions,before

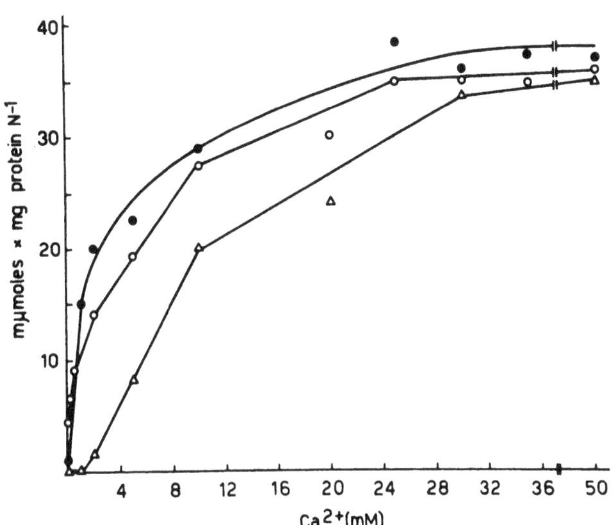

Fig.3 - The Effect of Adding Calcium Ions on the Incorporation of Labelled Serine into the Phospholipids of Brain Microsomes. Incubation has been carried out for 30 min in the standard incubation system described in Figs.1 and 2, except that calcium concentration was varied, as shown.
Activity was determined as reported in the text, and is expressed in nmoles of serine incorporated/mg of microsomal protein N/30 min.
Black circles : brain microsomes prepared in the sucrose-beta-mercaptoethanol-EDTA solution, washed in the same medium without EDTA, and again sedimented ;
White circles : brain microsomes prepared in the sucrose-beta-mercaptoethanol solution ;
Triangles : brain microsomes prepared as above, and incubated in the presence of 5 mM-EDTA.

the addition of L-serine and of increasing amounts of Ca^{++} ion. Here again an incorporating activity comparable to standard values (i.e., microsomes prepared in an EDTA-free medium) is reached only at a Ca^{++} ion concentration of about 10 mM.

The exchange mechanism for ethanolamine incorporation into lipid also requires calcium ions for optimum activity. The highest rate of incorporation is reached at 10 mM-Ca^{++} ion, and does not change with increasing amounts. This result compares well with the findings reported by CRONE (5) in his studies with housefly fat bodies, and by BORKENHAGEN *et al.* (2), who worked with rat liver homogenates, but is at variance from those of VANDOR and RICHARDSON (6), who found optimal incorporation at 2.9 mM-Ca^{++} ion, and an inhibition at higher levels. It might be well that different concentrations of substrates used might be at play in explaining these apparent discrepancies, by affecting the response to Ca^{++} ion.

All these results show the crucial requirement for Ca^{++} ion in the incubation medium and explain that the small incorporating activity exerted by the microsomes in the absence of added ions is only due to the presence of residual endogenous Ca^{++} ion in the particles.

EDTA acts by virtue of its complexing capacity, as shown further in Fig.4, which indicates that the pre-incubation of the microsomes with increasing amounts of EDTA gradually decreases the rate of serine uptake into lipid, when Ca^{++} ion is added. This effect is similar, in terms of relative activities, by adding either 5 mM or 25 mM calcium ions (Fig.4), thus showing that EDTA merely acts by complexing these cations.

The Ca^{++}-mediated incorporation of labelled serine into microsomal lipid is very slightly inhibited, at any Ca^{++} ion concentration, by the presence of Mg^{++} ion up to 40 mM. However, when a pre-incubation for 20 min of the microsomes is carried out with increasing amounts of Mg^{++} ion and L-serine (2.4 mM),before the addition of 25 mM $CaCl_2$, then a gradual reducftion in the existing enzymic activity is observable (Fig.5).

The incorporation of ethanolamine into microsomal lipid is inhibited by 20 % by the presence of 20-40 mM-Mg^{++} ion, as observed with parallel experiments. A higher inhibition was found by CRONE (5), but in the absence of added Ca^{++} ion.

Effect of other Metal Ions on the Enzymic Activity

The incorporation of serine and ethanolamine is actively inhibited, although at different degrees of inhibition, by Cu^{++}, Hg^{++}, Co^{++} and Mn^{++} ions. Pre-incubation of the microsomes with the metal ions is not necessary to achieve the inhibitory effects. The highest inhibition is produced by Cu^{++} ion, while the lowest is exerted by Co^{++} ion. Manganese ions are highly effective inhibitors of

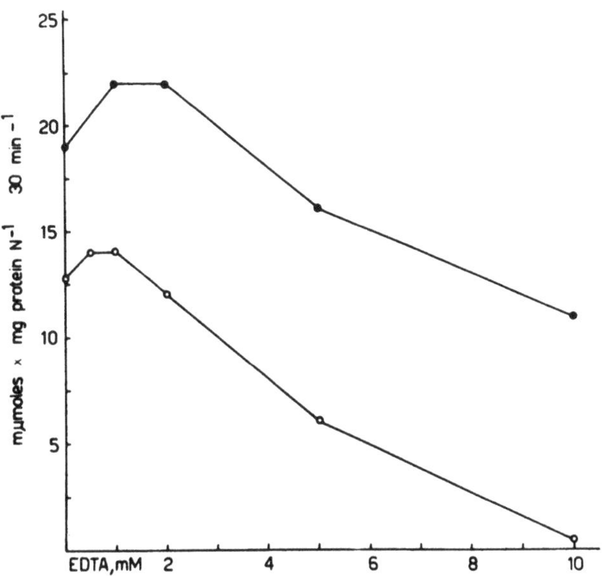

Fig.4 - The Effect of the EDTA Pre-Incubation on the Incorporation of Labelled Serine into the Phospholipids of Brain Microsomes.
White circles : microsomes, prepared in the sucrose-beta-mercapto-ethanol solution, were pre-incubated for 15 min with increasing amounts of EDTA, before the addition of L-serine and 5 mM $CaCl_2$, in the standard incubation system described in Figs.1 and 2 ;
Black circles : same as above, except that the $CaCl_2$ was 25 mM.

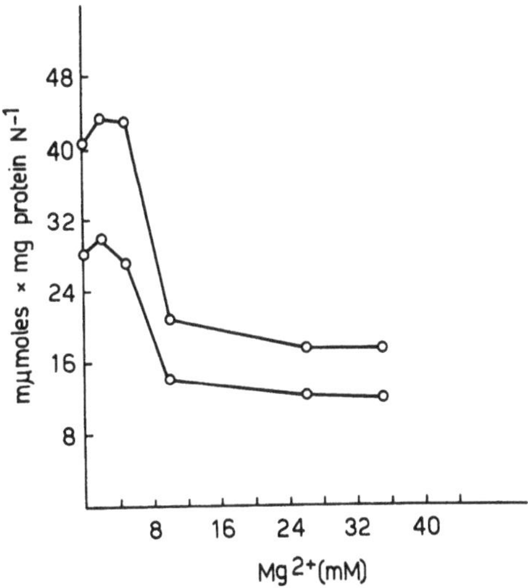

Fig.5 - The Effect of Mg^{++} Ion on the Ca^{++}-Stimulated Incorporation of Serine. The two experiments were done in similar conditions.

ethanolamine incorporation, as reported by CRONE (5) and HÜBSCHER (3).

Products of Incorporation

Analyses carried out on the total lipid extract, either by hydrolytic procedures (17), as described elsewhere (12-14) or by direct thin-layer chromatography (12,13,18) indicate that up to 85 % of the incorporated radioactivity after the serine exchange is recovered in diacyl sn-glycero-3-phosphorylserine (GPS), while a small percentage (8%) is found in diacyl sn-glycero-3-phosphorylethanolamine (GPE). The percentage of diacyl GPE formed is lower, if the incubation is reduced to 15 min, and derives presumably from the enzymic decarboxylation of the labelled diacyl GPS. Similar distribution of radioactivity have been reported by McMURRAY (7) and PORCELLATI, di JESO and MALCOVATI (8), who have worked also with brain preparations, and have been confirmed by us also with the use of thin-layer chromatography (TLC) procedures, by using various solvents.

It is known that serine is a very difficult lipid precursor to study, specially for in vivo experiments, since it enters into many metabolic pathways. By use of hydrolytic procedures we have however obtained evidence that the original activity present in the isolated diacyl GPS is almost exclusively localized in its hydrolytic product GPS, thus showing that the label is confined chiefly in the base moiety of the lipid, at least for the time period studied.

The ethanolamine incorporated into chick brain microsomal membranes was found to be associated with the ethanolamine phosphoglyceride (EPG) components. A large proportion (93 %) of the total incorporated radioactivity is recovered in the diacyl GPE, as it has been revealed by TLC. A small but significant amount (5%) is recovered in the ethanolamine plasmalogen (alkenyl acyl GPE), while no label is detectable in the choline-containing lipids (CPG). Experiments of hydrolysis of the labelled phospholipid derived from ethanolamine have also demonstrated that the radioactivity is entirely localized in the base-component of the diacyl GPE.

Inhibition Studies

With the use of inhibition experiments it has been observed in the course of our studies that L-serine and ethanolamine compete for the same enzymic site. We have revealed in fact that L-3-^{3}H-serine is effectively displaced from the enzymic site by 1 mM unlabelled ethanolamine. Similar results have been reported by BORKENHAGEN et al. (2) and CRONE (5), for rat liver homogenates and housefly fat bodies, respectively. The incorporation of labelled L-serine, on the other hand, is unaffected by choline, inositol and carnitine. These results partially confirm the findings of HÜBSCHER (3) and CRONE (5), but are at variance from those of CRONE himself, as regard to the effect produced by choline. D-serine (0.5-2.0 mM range) inhibits only slightly the incorporation of labelled L-serine, and is itself only feebly converted into lipid material. This result is similar to that of HÜBSCHER (3) and of CRONE (5).

As regard to ethanolamine, its incorporation into diacyl GPE is inhibited by L-serine, slightly affected by choline and unaffected by D-serine, inositol and carnitine (Table II). The inhibition caused by unlabelled L-serine is not due to its decarboxylation to unlabelled ethanolamine, which could dilute the pool of labelled 1,2-^{14}C-ethanolamine. In control experiments we have been unable in fact to detect any decarboxylation of 1-^{14}C-DL-serine with chick brain microsomes.

TABLE II

The Inhibition of 1,2-^{14}C-Ethanolamine Incorporation into the Chick Brain Microsomes

Compound added to the incubation system	Activity (% of control value)
2 mM-L-serine	31
2 mM-D-serine	98
2 mM-choline	82
2 mM-inositol	104
10 mM-inositol	95
4 mM-DL-carnitine	94
20 mM-DL-carnitine	96

The reaction mixture contained in the standard system 1.5 mM-1,2-^{14}C-ethanolamine (Specific Activity of 0.0375 microcuries/micromole) and inhibitors, as indicated. Results are presented as the radioactivity incorporated into diacyl GPE, expressed as a percentage of the control (15.6 nmoles/mg protein/hr) with no added amino alcohols.

Since the experiments presented in Table II have been done at one inhibitor (serine) concentration, it is impossible to decide whether the effects exerted by L-serine are due to competition or to a non-competitive inhibitions. However, we have already mentioned previously that ethanolamine inhibits competitively the incorporation of L-serine into lipid. Moreover, it was already reported by BORKENHAGEN et al. (2), MIEDEMA and RICHARDSON (19) and CRONE (5) that the inhibition of the ethanolamine incorporation by L-serine is of a competitive nature.

Choline and D-serine, not only do not displace or displace very slightly ethanolamine from the enzyme, but are themselves only feebly converted to lipid material under similar conditions.

Mechanism of Serine and Ethanolamine Incorporation

As we have reported in the present work, the chick brain microsomes incorporate into lipid, via a Ca^{++}-stimulated exchange, L-serine, as well as ethanolamine, but apparently not choline (unpublished observations). Owing to the different experimental conditions adopted (see Table III), the amount of incorporated ethanolamine is much lower than that of serine (about one fourth). Now, studies have been carried out in order to find out what the extent of labelling of lipid would be, if both labelled serine and ethanolamine were incubated together, in the same concentrations when incubated separately. Table III shows that, after 30 min of incubation of the two precursors, the rate of incorporation of the label into lipid averages between the two mentioned values, being approximately half the value obtained when serine alone is employed. No such effect is observable when labelled Me-^{14}C-choline is incubated together with the radioactive L-serine.

Subsequent experiments have been carried out with the aim of determining the ability of increasing amounts of non-radioactive L-serine to displace the radioactivity from microsomes containing diacyl GPE pre-labelled with 1,2-^{14}C-ethanolamine in vivo. Chick of 7 days of age were therefore injected intraperitoneally with 1,2-^{14}C-ethanolamine (0.1 microcurie/g), and then sacrified at the 6th day from the administration. Brain microsomes were prepared as usually, the specific activity of the endogenously labelled diacyl GPE determined by TLC, and additional aliquots of the fractions

TABLE III

The Effect of Combined Incubations of L-3-^{14}C-Serine and of 1,2-^{14}C-Ethanolamine with Chick Brain Microsomes *in Vitro*

Expt. No.	Precursor	Activity
1	3-^{14}C-serine	21,400
2	3-^{14}C-serine	19,850
3	1,2-^{14}C-ethanolamine	5960
4	1,2-^{14}C-ethanolamine	6170
5	Serine + ethanolamine	14,300
6	Serine + ethanolamine	12,780

For the experiments n.1 and 2, standard experimental conditions were adopted (Figs.1 and 2), except that L-3-^{14}C-serine (2.4 mM, 804 nc, S.A. of 0.67 microcuries/micromole) was used in place of the tritiated precursor. For the experiments n.3 and 4, conditions were similar to those reported for 1 and 2, except that 1,2-^{14}C-ethanolamine (2.27 mM, 170 nc, S.A. of 0.15 microcuries/micromole) was used in place of the labelled serine. In the experiments n.5 and 6 the two substrates were incubated together. Incubations were carried out for 60 min at 37°. Activity is expressed as total counts/mg of protein/60 min.

TABLE IV

The Displacement of Radioactivity from Diacyl GPE of *in Vivo*-Labelled Chick Brain Microsomes by Unlabelled L-Serine or Ethanolamine

Expt. no.	Addition	S.A. of diacyl 1,2-^{14}C-GPE
1	none	6.7
2	none	6.6
3	L-serine	2.7
4	L-serine	4.2
5	ethanolamine	4.4
6	ethanolamine	3.9

The microsomal preparation (see the text) was incubated in standard conditions (see Figs.1 and 2) without substrates (expt.1 and 2), and with 2.4 mM unlabelled L-serine (expt.3 and 4) or 2.27 mM unlabelled ethanolamine (expt.5 and 6). The approximate final concentration of the microsomal diacyl GPE was 0.6 mM. S.A. (nc/micromole) at zero time was between 6.9 and 7.4.

(corresponding to about 200–250 micrograms of protein N) were incubated *in vitro* with unlabelled L-serine or ethanolamine under standard experimental conditions. After incubation, the diacyl GPE was separated and its specific activity again determined. The results of Table IV show a decrease of the specific activity of the labelled diacyl GPE, when the radioactive microsomes are incubated in the presence of either unlabelled L-serine or ethanolamine.

These results therefore indicate a release of the labelled ethanolamine from the endogenous diacyl-1,2-^{14}C-GPE, with a parallel exchange with external L-serine or ethanolamine, thus providing evidence for postulating that the incorporations of 3-^{14}C-serine and of 1,2-^{14}C-ethanolamine represent an exchange of the labelled free base with the bound base of diacyl GPE (and probably of other lipids). In other words, unlabelled ethanolamine or serine, in addition to inhibiting the rate of incorporation of labelled L-serine or labelled ethanolamine, respectively, also exchange with the 1,2-^{14}C-ethanolamine already present as diacyl-1,2-^{14}C-GPE, thus producing a decrease in the previously incorporated radioactivity.

The Effect of Adding Phospholipid on the Enzymic Activity

The preceding sections have emphasized the fact that chick brain microsomal membranes carry out an exchange reaction between free bases and their endogenous phospholipids. Although a variety of phospholipids might actually exchange with the bases *in vivo* and although it is known that phospholipids of particular makeup exist within biomembranes, which may be highly important for correct interactions with structural and catalytic proteins so as to form membranes, it was decided to carry out a few experiments in order to show the effect of adding some phospholipid classes on the enzymic exchange.

With preliminary experiments it was shown that an almost complete recovery (ranging from 95 to 100 %) of endogenous lipid material and of externally added phospholipid, as well as of radioactivity, was obtained from the precipitate, by treating the incubation mixture and by determining the incorporating activity, as reported in Fig.1.

Table V shows that adding suitable levels of bovine diacyl GPS to a minimal amount of microsomal membranes (about 0.2 mg of microsomal protein, corresponding to 25 mg of original fresh tissue), increases the incorporating activity to some degree ; however, the increase levels off by using higher phospholipid concentration. Similar results have been obtained, by using bovine diacyl GPE, in place of the serine derivative, under similar conditions.

TABLE V

The Effect of Adding Phospholipids upon the Exchange Reaction of L-Serine in Chick Brain Microsomes *in Vitro*

Addition (final concentration)	Activity (a)	(b)	(c)	(d)
none	6.4	9.3	8.0	9.2
0.2 mM diacyl GPS	8.3	11.0	9.1	-
0.4 mM diacyl GPS	10.5	11.4	9.1	4.5
0.8 mM diacyl GPS	7.6	11.7	8.6	-
1.2 mM diacyl GPS	7.6	11.4	7.4	-
1.5 mM diacyl GPS	5.9	8.1	7.9	-

Standard incubation mixture and experimental conditions, as reported in Figs.1 and 2. In the experiments (a), diacyl GPS (5 mM) was sonicated in a volume of 10 ml for 30 min in a 100 Watt MSE ultrasonic disintegrator (100 W, 20 Kcycles, 8 microns peak-to-peak on amplitude meter) in a 0.32 M-sucrose-40 mM-Veronal buffer (pH 8.5)-4 mM-dithiothreitol (DTT) solution at +2°, before adding. The sonication treatment was carried out for periods of 30-45 sec, followed by 30 sec of standing. In the experiments (b), diacyl GPS was sonicated under similar conditions, but by omitting the Veronal buffer and DTT. In the experiments (c), bovine serum albumin (0.2 % of final concentration) was added to the mixture during the sonication. In the experiments (d), the microsomal suspension was sonicated together with the previously sonicated diacyl GPS for a few seconds, before incubating.
Incorporation rates were determined as described in Fig.1. Activity is expressed as nmoles/mg of protein/30 min.

It is impossible to decide by these experiments whether there occurs a substrate inhibition of the enzyme or purely physical factors may be at play in explaining this substrate-velocity curve relationship, by interfering with the compenetration of the lipid phase (diacyl GPS or diacyl GPE) with the enzyme (microsomal membrane). Moreover, the endogenous phospholipid content of the microsomes, although very low, certainly will influence the reaction rate.

It is interesting to note from Table V that the sonication of the microsomal membranes, carried out together with the added diacyl GPS before incubation, lowers the incorporating activity by some degree.

The gas chromatographic analyses performed on the two phospholipid classes used in the experiments of Table V (Appl.Sci.Labs., Inc., State College, Penn., U.S.A.) have revealed the following fatty acid composition (percentage values between brackets) : diacyl GPS : 16:0 (3.2), 18:0 (32.7), 18:1-18:2-18:3 (51.6), 20:3-20:4 (7.1), 22:5-22:6 (5.4) ; diacyl GPE : 16:0 (9.7), 16:1 (2.3), 18:0 (9.7), 18:1 (34.7), 20:0 (1.5), 20:3-20:4 (15.1), 22:5-22:6 (24.4), 24:1 (1.7). Of course, it is not known what effect the different fatty acid composition, i.e. the different molecular species of the diacyl GPS or of the diacyl GPE could exert upon the exchanging capacity of the brain microsomes. An experimental approach to this problem has been previously undertaken by several Authors for other purposes, and has been also provisionally examined by us by incubating the microsomal membranes (25 mg of fresh starting material) and the labelled L-serine with added microsomal phospholipid preparations, obtained by preparative TLC of the same membranes, and by comparing the results with those obtained by the use of a mixture of standard phosphlipid classes, incubated in about the same proportions as those which were seen to exist in the isolated microsomal membranes.
Table VI shows, in this connection, that a definite, although slightly higher, effect is exerted by the natural phospholipids, as compared to that brought about by phospholipid material of different source. This result might signify that the fatty acid composition of the exchanging phospholipid molecules does effectively influence the activity of the enzymic system.

The Effect of Adding Detergents on the Enzyme Activity

With an other set of experiments, it was decided to carry out solubilization treatments of the enzymic system, in order to obtain a material as much as possibly free from endogenous phospholipid.

TABLE VI

The Effect of Adding Microsomal Phospholipid on the Exchange of L-Serine in Chick Brain Microsomes in Vitro

Expt. No.	Addition	Activity
1	none	9.2
		8.5
2	Microsomal Phospholipid*	15.7
		16.6
		17.1
		15.5
3	Non-microsomal Phospholipid**	12.6
		11.0
		12.2
		11.8

Experiments were performed with the microsomes corresponding to 25 mg of fresh weight of tissue (about 0.2 mg of protein), as the enzymic source, incubated in standard conditions at 37° for 30 min. Activity is expressed as nmoles of serine incorporated/mg protein/ 30 min.

* The microsomal phospholipid were prepared according to FISCUS and SCHNEIDER (21) from chick brain microsomes, and incubated under similar conditions. They were represented for 100 mg of original fresh tissue by 0.42 micromole of diacyl GPC, 0.20 of diacyl GPE, 0.14 of diacyl GPS, 0.05 of diacyl GPI and 0.08 of sphingomyelin. A final concentration of these total microsomal lipids of 0.9 mM was used in a final volume of 0.5 ml.

**Incubation was carried out with phospholipids from Appl.Sci.Labs., Inc., U.S.A. Final concentration of 0.9 mM. Lipid classes employed were the following : diacyl GPC (about 0.2 micromole), diacyl GPE (0.1), diacyl GPS (0.08), diacyl GPI (about 0.03) and sphingomyelin (about 0.05). Final volume of 0.5 ml.

Since a treatment with detergents was choosen, it was decided to examine first what the behavior of the enzymic system would have been in the presence of definite levels of some of the most commune detergents and what would have been the effect of pre-incubating the microsomal membranes with these compounds.

Table VII shows that a strong inhibitory effect on the serine exchange is brought about by the various detergents which have been used, although different degrees of inhibition were obtained ; however, pre-incubating the microsomal membranes with these detergents for periods of 30 min, before incubating with Ca^{++} ion, produces much slighter inhibitory effect on the enzymic system, so that it can be observed that sodium cholate (0.1%), Tween 2o, Cutscum and sodium deoxycholate nearly do not inhibit the exchange, when pre-incubated with the enzymic source. These results can probably be explained, in view of the fact that some detergents may in some way unmask the pre-existing enzymic activity (thus "evidentiating" a higher exchange rate), and that in relative terms this effect could be higher or more rapid than inhibiting the same enzymic system, so that on the whole no great inhibitory action is revealed in these conditions by some detergents as compared to the strong effect exerted by the same compounds in the non-preincubated experiments (Table VII, a). The absence of Ca^{++} ion during the pre-incubation treatment may also be important, in this connection, in explaining the failure of some detergents of inhibiting the enzymic exchange.

TABLE VII

The Effect of Adding Some Detergents upon the Membrane-Bound Enzymic Activity of Chick Brain Microsomes

Addition	Final Concentration (%)	Activity (a)	(b)
None	-	0.33	0.30
Sodium cholate	0.1	0.13	0.26
Sodium cholate	3.0*	0.14	-
Tween 20	0.1	0.22	0.28
Cutscum (Fischer)	0.1**	0.07	0.24
Sodium deoxycholate	0.1**	0.11	0.23
G 3634 A (Atlas)	0.1	0.09	0.04
Triton X-100	0.1**	0.04	0.02

The experiments (a) were done without pre-incubating the detergents with the microsomes. The experiments (b) were done by pre-incubating the microsomes for 30 min in the standard system without Ca^{++} ; after that time, Ca^{++} ion was added and normal incubation carried out for 30 min. Activity expressed as nmoles/min/mg protein N.

* Complete solubilization.

**Almost complete solubilization.

Because of the fact that sodium cholate, a widely used solubilizing agent, does not produce high inhibitory effects upon the exchange reaction between serine and phospholipid, while producing a clear clarifying action upon the microsomal suspension, when used in the amounts of 0.5 % of final concentration, it was decided to use this detergent in the subsequent experiments, planned to try to solubilize the enzymic activity.

Experiments with Solubilizing Procedures

Before carrying out this type of experiments, it was demonstrated that an almost complete recovery of endogenous lipid material and of externally added phosholipids, as well as of radioactivity, was reached in the presence of sodium cholate in the experimental design (0.5 % of final concentration).

Different treatments of the microsomal membranes were performed in the presence of sodium cholate, with the aim of solubilizing the enzymic system and of recovering the activity, with and without the addition of suitable amounts of phospholipids.

None of the procedures used has yielded preparations with measurable enzymic activity. After the various treatments which have been choosen, exchange activity was invariably lost, and could not be restored or only slightly restored, upon addition to the treated system of suitable amounts of various phospholipid classes.

Despite these results, a lyophilization treatment of freshly prepared microsomal membranes, followed by an ultrafiltration procedure at pH 10.0 and subsequent washing at pH 8.5, as reported in Table VIII, yielded much better results. It appears from the Table that, although a great portion of the exchanging activity is lost by the complete treatment of the brain microsomes (and not by the simple lyophilization), the enzymic activity is fully recovered, with more than a two-fold increase, upon the addition to the treated material of 0.4 mM diacyl GPS. Contemporarily, the content of the endogenous phospholipid in the treated membranes drops by more than 75 %, thus allowing us to work with an enzymic system noticeably devoid of the bulk of the endogenous phospholipid content, which certainly disturbs further experiments. It is worthy mentioning that the feasibility of such a solubilizing and delipidizing procedure could allow us to use other detergents or other experimental variations, in order to improve the final results.

TABLE VIII

Serine Exchange Activity, Phospholipid Content and Protein Recovery after Lyophilization and Ultrafiltration of Chick Brain Microsomes

	Activity*	Protein**	Microsomal Lipid°
Control Membranes	0.50	802	0.950
	0.86	806	0.964
	0.66	785	0.932
	0.47	735	0.876
Lyophilized Membranes	0.59	740	0.851
	0.68	812	0.941
Treated Membranes^	0.25	780	0.203
	0.29	765	0.231
	0.22	743	0.198
	0.56°°	784	0.210
	0.73°°	731	0.187
	0.51°°	751	0.208

^ The microsomal suspension in sucrose (0.32 M) was lyophilized, and then suspended in 2 mM DDT. Ultrafiltration was carried out on an "Amicon" centriflo-type ultrafilter, after having supplemented the suspension with sodium cholate (3 % final concentration), and bicarbonate buffer, pH 10.0 (20 mM). After this treatment, the ultrafilter content was washed once with sucrose (0.32 M), DTT (2 mM) and Tris-HCl buffer, pH 8.5 (40 mM), and again ultrafiltered. The amount of 100 mg fresh tissue was then supplemented with the incubation components, and incubated, as usual.

* nmoles/mg protein N/min.

** micrograms/100 mg wet wt.

° micromoles/100 mg wet wt.

°° 0.4 mM bovine diacyl GPS (Appl.Sci.Labs., Inc., U.S.A.), prepared as described in Table V, was added to the incubation mixture.

GENERAL CONSIDERATIONS

The Ca^{++}-stimulated serine incorporation reported in this study is probably the only way by which this precursor is converted into lipid in brain. The conversion takes place in the absence of any added energy-source and gives rise to a labelled lipid, diacyl GPS, whose radioactivity is almost completely confined to its base-component for the time-periods studied. Similarly, a calcium-stimulated system occurs in chick brain which incorporates enzymically labelled ethanolamine into phospholipids, and which is similar in many aspects to the serine-exchange system, and to the ethanolamine-incorporating system described in liver tissue (2), housefly fat bodies (5) and plants (6). The system has no relation at all with the cytidine-dependent pathway of EPG synthesis, because of the various reasons which have been reported under RESULTS.

It is interesting to note that, if we consider the mean value of microsomal diacyl content of 7 days-old chick brains, which we have estimated to be 0.78 ± 0.12 micromoles/g of original fresh tissue (13 assays), and if we examine the results of the present work, then we may conclude that the Ca^{++}-stimulated exchange mechanism for diacyl GPS biosynthesis nearly increases by a value of about 15 % the original diacyl GPS content of the brain microsomes. On the other hand, the Ca^{++}-stimulated exchange mechanism for the incorporation of ethanolamine into EPG <u>in vitro</u> accounts for a rate of 220 nmoles/g of original fresh tissue/hr, which represents about 5 % of the incubated ethanolamine in our conditions. If we consider the mean value of the microsomal diacyl GPE content of 7 days-old chick brains, which we have shown to be about 1.4 micromoles/g of original fresh tissue (see 12), then we may conclude that the Ca^{++}-dependent exchange mechanism for diacyl GPE biosynthesis increases by a value of about 15 % the original diacyl GPE content, as it was the case of the diacyl GPS.

The enzymic system(s) which incorporate(s) serine and ethanolamine have similar characteristics (similar subcellular distribution patterns, very close K_m values, same inhibitory effect produced by the heavy metal ions, similar pH optimum) ; some minor apparent differences existing between the two systems are not real, in our opinion, when closely examined. They can be explained in terms of different affinities of the nitrogenous compound for Ca^{++} and Mg^{++} ions or for the Tris buffer, rather than to differences in possible enzymic systems. These and many other considerations, as well as the results about the competitive inhibition produced by ethanolamine on serine incorporation and viceversa, let us suppose that the incorporation of labelled L-serine and ethanolamine (and probably of other amino-alcohols) by the chick brain microsomal fraction is catalyzed by a single enzyme system and not by different enzyme actions, due to the reversal of phospholipase D (E.C. 3.1.4.4) activity (1,3). This consideration, which has been put forward in other occasions (2,5,6), is substantiated by the following findings and considerations : (a), phospholipase D has never been reported to occur in animal tissue (see 21) ; (b), the addition of cabbage phospholipase D, phosphatidic acid or diglyceride does not produce an increase of the rate of serine and ethanolamine incorporation into lipid under basic experimental conditions ; (c), ethanolamine and L-serine compete for the same enzymic site, and in addition an intermediate rate of conversion to lipid is obtained when both substrates are employed together, thus suggesting a single mechanism of incorporation ; (d), the radioactivity content of endogenously labelled diacyl GPE decreases on incubating unlabelled L-serine or ethanolamine ; (e), incorporation is stimulated by the presence of added phospholipids.

As regard to these last compounds, the presence of endogenous phospholipid in the enzymic source (isolated microsomes) used in this work, is certainly disturbing in carrying out kinetic studies and in examining the lablling in various lipid classes and from a variety of labelled precursors. Ideally, the best conditions to study such exchange reactions would be to obtain a lipid-free enzyme preparation. In this connection, we have shown in this study some results about an enzymic preparation obtained from the microsomal membranes by use of lyophilization and detergents, which is devoid of the bulk of the phospholipid, but experiments must be improved because of the low yield of activity reached in these conditions. Only by using lipid-devoid microsomal enzyme it could be possible in the future to examine the physiological significance and the quantitative efficiency of these calcium-stimulated reactions.

ACKNOWLEDGEMENTS

The work was supported in part from the C.N.R., Rome (Contract n.69.02216.115.3381.0). Thanks are given to Dr.R.E.McCaman (Duarte, Ca., U.S.A.) for the supply of some products mentioned in the work.

REFERENCES

1. Dils, R.R. and Hübscher, G., Biochim.Biophys.Acta 46:505 (1961).

2. Borkenhagen, L.F., Kennedy, E.P. and Fielding, L., J.Biol.Chem. 236:PC 28 (1961).

3. Hübscher, G., Biochim.Biophys.Acta 57:555 (1962).

4. Artom, C. and Weiner, A., Feder.Proc. 22:415 (1963).

5. Crone, H.D., Biochem.J. 104:695 (1967).

6. Vandor, S.L. and Richardson, K.E., Canad.J.Biochem. 46:1309 (1968).

7. McMurray, W.C., J.Neurochem. 11:287 (1964).

8. Porcellati, G., di Jeso, F. and Malcovati, M., Life Sci. 5:769 (1966).

9. Ansell, G.B. and Spanner, S., Biochem.J. 100:50 P (1966).

10. Pirotta, M., Giorgini, D. and Porcellati, G., Boll.Soc.It.Biol. Sper. 45, issue n.20 bis, communic. n.208, (1969).

11. Arienti, G., Pirotta, M., Giorgini, D. and Porcellati, G.,Biochem. J. 118:3 P (1970).

12. Porcellati, G., Biasion, M.G. and Arienti, G., Lipids 5, in the press (1970).

13. Porcellati, G., Biasion, M.G. and Pirotta, M., Lipids 5, in the press (1970).

14. Porcellati, G., Pirotta, M., Arienti, G. and Giorgini, D., J. Neurochem., in the press (1970).

15. Folch-Pi, J., Ascoli, I., Lees, M., Meath, J.A. and LeBaron, F.M., J.Biol.Chem. 191:833 (1951).

16. Folch-Pi, J., Lees, M. and Sloane-Stanley, G.G., J.Bil.Chem. 226: 497 (1957).

17. Dawson, R.M.C., Hemington, N. and Davenport, J.B., Biochem.J. 84:497 (1962).

18. Cuzner, M.L. and Davison, A.N., J.Chromat. 27:388 (1967).

19. Miedema, E. and Richardson, K.E., Plant Physiol. 41:1026 (1966).

20. Fiscus, W.G., and Schneider, W.C., J.Biol.Chem. 241:3324 (1966).

21. Björnstad, P., Biochim.Biophys.Acta 116:500 (1966).

MEMBRANE-BOUND HYDROXYMETHYLGLUTARYL COENZYME A REDUCTASE.

B. Hamprecht, K.R. Bruckdorfer[1)], C. Nüßler

and F. Lynen

Max-Planck-Institut für Zellchemie, München

Cholesterol can be formed from acetate via a large number of enzymatic steps (Fig. 1). Some of the

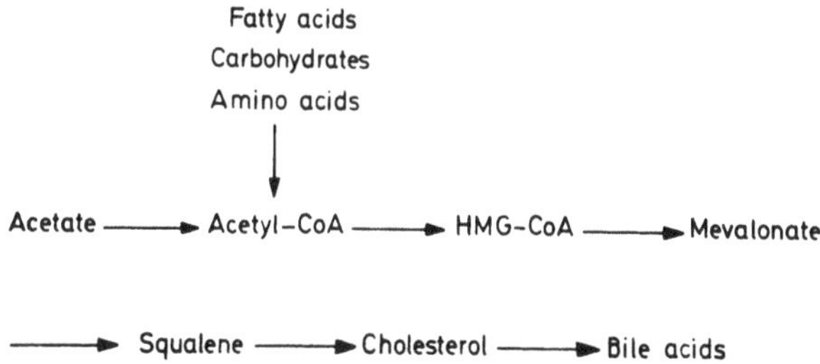

Fig. 1. Enzymatic steps from acetate to cholesterol

more important intermediates are acetyl-CoA, 3-hydroxy-3-methylglutaryl coenzyme A (HMG-CoA), mevalonate and squalene. The enzyme, which catalyzes the rate limiting reaction of this sequence is 3-hydroxy-3-methylglutaryl coenzyme A reductase (E.C. 1.1.1.34; HMG-CoA reductase). It enables the reduction of the thioester bonding in HMG-CoA to the primary alcohol function of mevalonate by using two molecules of NADPH as reductant (Fig. 2). This enzyme has first been

1) Present address: University of London, Queen Elizabeth College, Dept. of Nutrition.

discovered in yeast (1). As it catalyzes the slowest reaction of the whole sequence from acetyl coenzyme A to cholesterol, it is the object of intensive research in those groups, who are interested in the mechanism of regulation of hepatic cholesterogenesis.

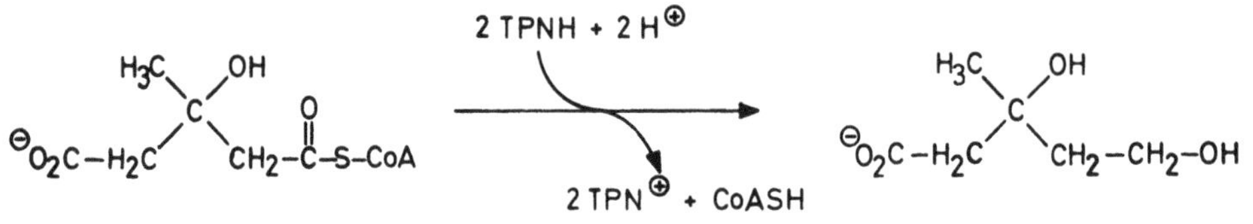

$$^{\ominus}O_2C-H_2C-C(CH_3)(OH)-CH_2-\overset{O}{\overset{\|}{C}}-S\text{-}CoA \xrightarrow[2\,TPN^{\oplus}\,+\,CoASH]{2\,TPNH\,+\,2\,H^{\oplus}} {}^{\ominus}O_2C-H_2C-C(CH_3)(OH)-CH_2-CH_2-OH$$

Fig. 2. Reaction catalyzed by HMG-CoA reductase.

The yeast enzyme has the advantage that one can obtain it readily in a solubilized form (1,2), whereas difficulties arise with the liver enzyme. The latter is a constituent of the microsomal fraction of a liver homogenate.

The activity of this enzyme in liver is extremely low and the microsomal fraction contains NADPH: cytochrome c oxidoreductase. Hence it is not possible to assay this enzyme by an optical method. Therefore, we developed a radiogaschromatographic assay procedure (3,4). 5-^{14}C-HMG-CoA is incubated with rat liver microsomes in the presence of a NADPH regenerating system. After incubating for 1 hour carrier mevalonate is added and the reaction is stopped by heating. The mixture is brought to dryness. After acidification the mevalolactone formed is extracted with ether. The residual material, after evaporation of the solvent is trimethylsilylated to form the trimethylsilylether of mevalolactone. An aliquot of this mixture is subjected to gaschromatography. The first peak in the mass diagram (Fig. 3a) is caused by tridecanoic acid methylester, an internal standard to calibrate the mass detector; the second peak corresponds to mevalolactone trimethylsilylether. The material emerging from the mass detector of the gaschromatograph is continuously combusted (5). The $^{14}CO_2$ formed is driven by the carrier gas through a flow-through scintillation cell filled with anthracene crystals, which is mounted in a scintillation spectrometer. This allows the integration of counts belonging to a peak and, at the same time, with a ratemeter the analog registration of a radioactivity diagram (Fig. 3b). The first group

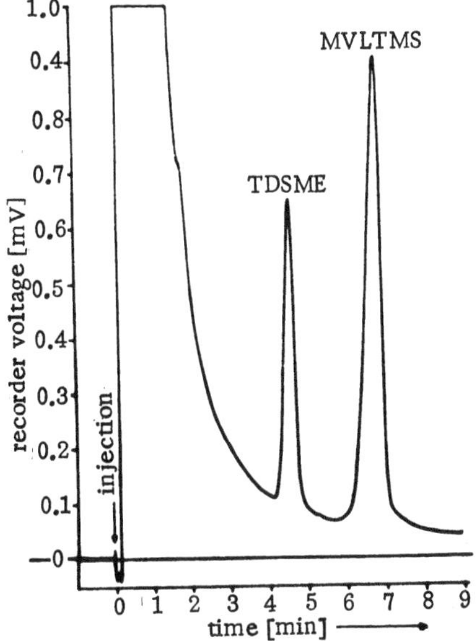

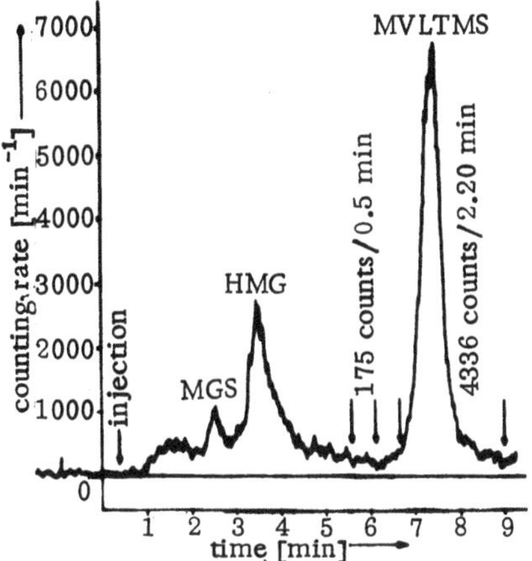

Fig. 3. a) Mass gaschromatogram for the determination of HMG-CoA reductase activity (upper diagram). A 50 μl aliquot of the solution of the trimethylsilylated ^{14}C-mevalolactone is injected into a Varian Aerograph 1520 gaschromatograph. Carrier gas He 25 ml/min; temperature of injector 250° C, of column oven 190° C, of detector oven 300° C. Columns 1/8 inch outer diameter, stainless steel, filled with 14 % neopentylglycol succinate on chromosorb W, acid washed, 80-100 mesh. TDSME = tridecanoic acid methylester, MVLTMS = mevalolactone trimethylsilylether.

b) Corresponding radioactivity diagram (lower diagram). For radiogaschromatography a Tricarb combustion furnace model 325, a Tricarb scintillation spectrometer model 3101, a ratemeter model 280 A, all from Packard Instrument, Frankfurt/Main, were used. The trimethylsilyl derivatives of cis- and trans-3-methylglutaconic acids are designated MGS, that of 3-hydroxy-3-methylglutaric acid is designated HMG.

of peaks is caused by the trimethylsilyl derivatives of cis- and trans-3-methylglutaconic acids and 3-hydroxy-3-methylglutaric acid. The second peak belongs to mevalolactone trimethylsilylether. From a large number of data taken from mass- and radioactivity measurements one can calculate the specific activity of the enzyme. As it is more convenient, a computer is used.

Within a limited range the activity of the enzyme is proportional to the amount of the protein used in the assay mixture (Fig. 4). The curve levels off be-

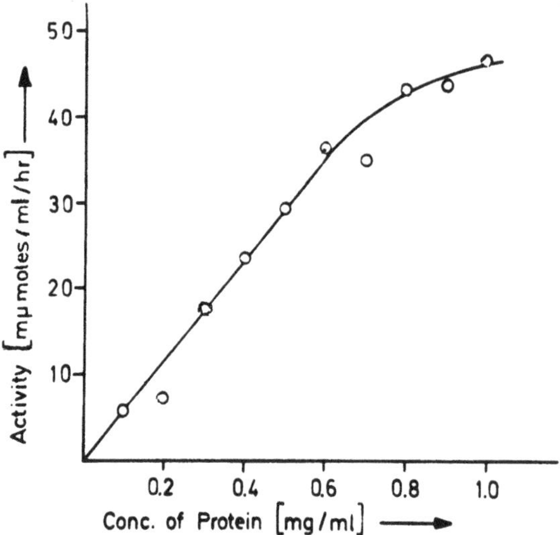

Fig. 4. Relationship between the concentration of microsomal protein in the assay mixture and the activity of HMG-CoA reductase.
The incubation mixture contained: Tris-HCl-buffer 100 mM, pH 7.2; K-EDTA 20 mM, pH 7.2; cysteine 20 mM; glucose-6-P 10 mM, pH 7.2; NADP 1 mM; glucose-6-P dehydrogenase (E.C.1.1.1.49) 10 μg/ml (110 units/mg); 0.075 mM 5-^{14}C-HMG-CoA (Spec. radioactivity 5.1x10^{7} dpm/μmole); varying amounts of microsomes prepared according to Regen et al. (6). The volume was made up to 0.200 ml by addition of homogenization buffer (6). Incubations were for 1 hr at 37° C.

cause the system becomes short of the substrate HMG-CoA.

The enzyme is only active in the presence of substances containing thiol groups (table 1). With other reductants only part of the total activity is found. In addition, the inhibition of the enzyme by iodoacetamide indicates that one or several SH-groups are essential for activity (Fig. 5).

TABLE I

Influence of Different Reductants on the Activity of HMG-CoA Reductase

Expt. No.	Reductant	Concn. mM	Spec. activity nmoles/mg/hr	Activity % of expt.1
1	cysteine	20	12.3	100
2	glutathione	20	14.9	121
3	dithiothreitol	20	12.2	99
4	thioglycol	20	14.1	115
5	ascorbate	20	0.3	2
6	KBH_4	5	0.7	6
7	$S_2O_4^{2-}$	10	7.1	58
8	$S_2O_5^{2-}$	10	8.8	72

The reductants were added immediately before starting the enzyme assay. For incubation conditions see text with fig. 4. The concentration of microsomal protein was 2 mg/ml.

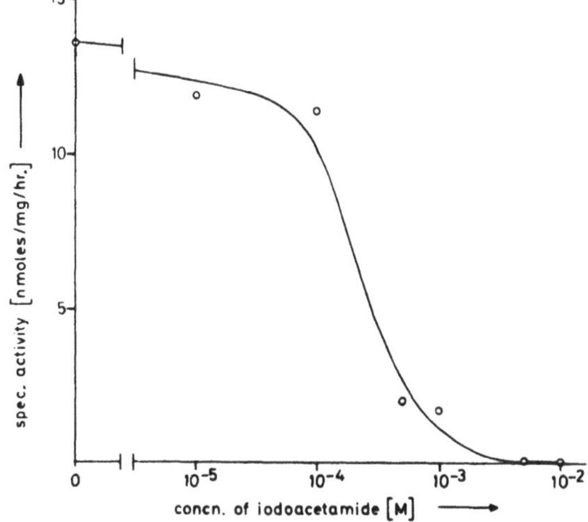

Fig. 5. Inhibition of HMG-CoA reductase by iodoacetamide. 0.070 ml 0.04 M potassium phosphate buffer, pH 7.2 and 5.6 $\times 10^{-x}$ or 2.8$\times 10^{-x}$ M iodoacetamide were added to 0.390 ml of a suspension of microsomes (16.9mg protein/ml), in order to obtain 0.560 ml of 1$\times 10^{-x}$ M solution of iodoacetamide. After 15 min of incubation at 25° C the reactions were stopped by addition of 0.100 ml of 0.1 M cysteine in 0.04 M potassium phosphate buffer, pH 7.2, containing 0.1 M sucrose and 0.05 M KCl. 0.040 ml of these mixtures were used to start the enzyme assay.

It is well known that the feeding of cholesterol to mammals results in a strong reduction of HMG-CoA reductase activity (7; literature cited there). To test the possibility that cholesterol might exert this effect on the level of the intact enzyme, we assayed microsomal HMG-CoA reductase in the presence of finely dispersed cholesterol (8). As expected, no inhibition could be found (table 2). The same result was obtained when cholesterol-lecithin micelles or lecithin alone were used (Fig. 6).

TABLE II

Influence of Cholesterol on HMG-CoA Reductase Activity

concn. of cholesterol mg/ml	specific activity nmoles/mg/hr experiment A	B
0.00	1.9	12.3
0.67	1.1	13.0
0.67	1.7	--

The incubations were performed as described in fig. 4, with the exception that a suspension of cholesterol was added where indicated. Concentration of microsomal protein: 2 mg/ml.

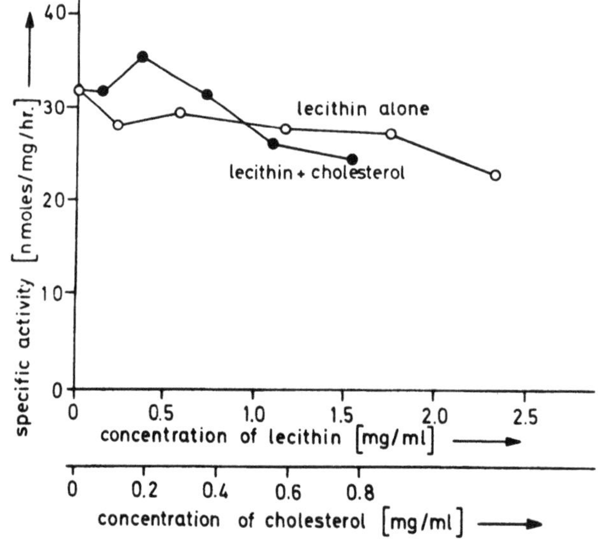

Fig. 6. Influence of lecithin/cholesterol micelles or lecithin alone on HMG-CoA reductase activity.

Lecithin was isolated from egg yolk (9). The micelles were prepared according to Bruckdorfer et al. (10). The molar ratio of lecithin:cholesterol was 0.98:1. The incubations were performed as described in fig. 4, except that the micelles were added. Protein concentration:1 mg/ml.

Feeding of bile acids also renders strong depression of HMG-CoA reductase activity in rat liver (11). On the basis of in vitro studies it has been suggested that bile acids are feedback inhibitors of cholesterol synthesis (12). Indeed, bile acids inhibit HMG-CoA reductase in vitro (Fig. 7). But this inhibition is exerted only in concentrations much higher than those found in vivo in the liver (13). To obtain insight into the mode of action of bile acids on HMG-CoA reductase in vitro the saturation curves of the enzyme for HMG-CoA in the presence and the absence of taurochenodeoxycholate were measured. The curves display no sigmoidal shape, even not when the inhibitor was present (Fig. 8).

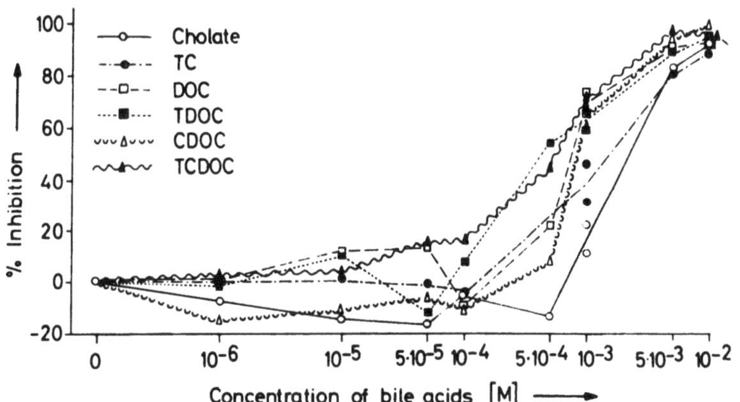

Fig. 7. Inhibition of HMG-CoA reductase activity by bile acids.
The incubations were performed as described in fig. 4, except that bile acids were added to the assay mixtures. Protein concentration:2 mg/ml.

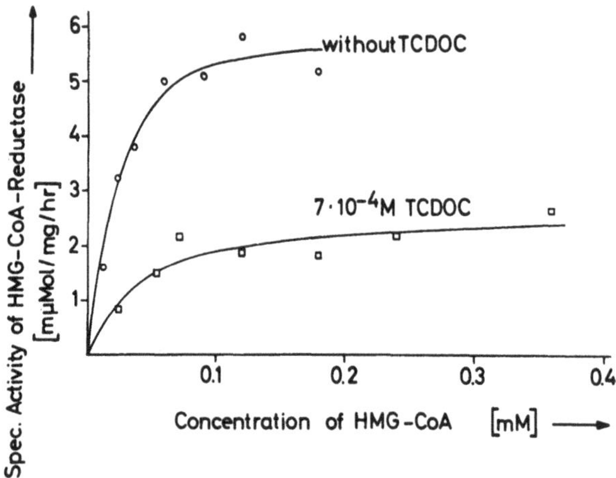

Fig. 8. Influence of HMG-CoA concentration on the activity of HMG-CoA reductase in the presence and the absence of taurochenodeoxycholate. Incubations were performed as described in fig. 4, except that the bile acid was added to some of the incubations and that the concentration of HMG-CoA was varied. Protein concentration : 2 mg/ml.

Sigmoidal shape of the saturation curve would be expected, if HMG-CoA reductase is an allosteric enzyme. The inhibition is noncompetitive, as seen from figures 8 and 9.

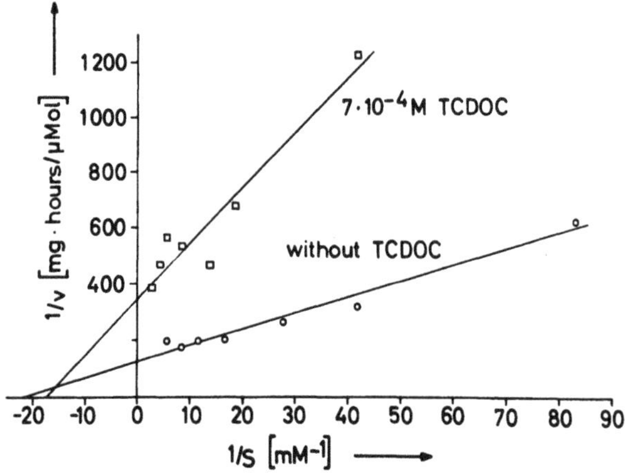

Fig. 9. Lineweaver-Burk plot of the data of fig. 8.

For an allosteric enzyme one would expect the inhibition to be totally reversible. In order to find out whether or not the inhibition of HMG-CoA reductase by bile acids is reversible, microsomes were preincubated with 10^{-3} M taurochenodeoxycholate. At different times after starting the incubation samples were taken for the assay of HMG-CoA reductase activity. In the assay mixture the concentration of the bile acid was diluted down to 10^{-4} M. Parallel to these, enzyme assays were performed in the absence of the bile acid or in the presence of 10^{-4} M or 10^{-3} M taurochenodeoxycholate, respectively. From fig. 10 it is seen that the activity of the enzyme declines with time. This may be due to enzyme denaturation. But it is not important in this connection. As expected there is no effect of 10^{-4}M bile acid, when only added to the enzyme assay (comparison of the upper two curves). On the other hand 10^{-3} M bile acid in the assay causes in-

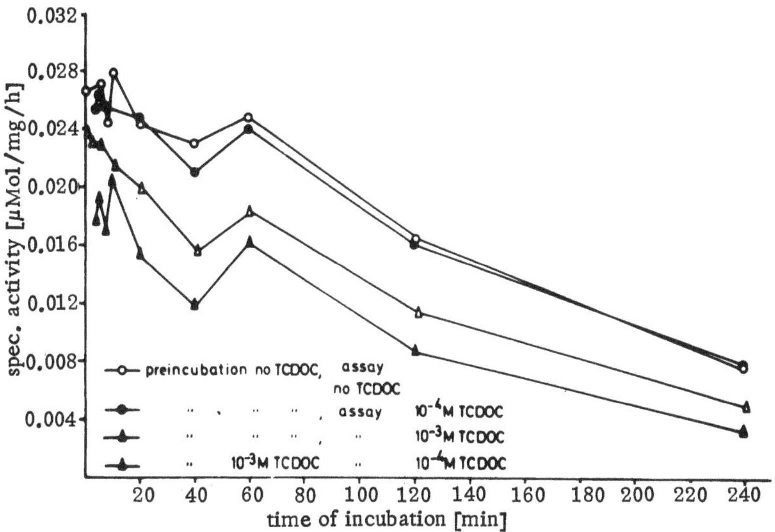

Fig. 10. On the reversibility of bile acid inhibition of HMG-CoA reductase. The preincubations were done at 37° C in the presence or the absence of 10^{-3} M taurochenodeoxycholate. After starting the preincubations, samples of 20 µl were taken at different times and assayed for enzyme activity, as described in fig.4. The protein concentration in the assay mixtures was 1 mg/ml. One series of assays with microsomes not pre-incubated with the bile acid was done in the absence of taurochenodeoxycholate, another in the presence of 10^{-4} M, a third with 10^{-3} M taurochenodeoxycholate, respectively.

hibition of the enzyme. If the inhibition caused by 10^{-3} M bile acid were fully reversible, the curve for the experiment with 10^{-3} M bile acid in the pre-incubation, but 10^{-4} M in the enzyme assay, should be congruent with the upper two curves. It should be congruent with the lowest curve, if the inhibition were fully irreversible. The experiment shows that the inhibition is only partially reversible (fig. 10). This precludes the possibility that taurochenodeoxycholate acts as an inhibitor of an allosteric enzyme.

The inhibition by bile acids of HMG-CoA reductase seems to be a nonspecific one, as has been observed with a number of other enzymes, too (14). This view is supported by the effect observed with other detergents (table 3). The nonionogenic detergent Triton WR-1339 does not influence the activity of the enzyme.

TABLE III

Effect of Different Nonionogenic Detergents on HMG-CoA Reductase

detergent	conc. %(w/v)	spec. activit of HMG-CoA reductase nmoles/mg protein/hr
none	-	45
Triton X-100	0.1	12
	1.0	1
Triton WR-1339	0.1	54
	1.0	49
Tween 20	0.1	43
	1.0	23
Tween 40	0.1	40
	1.0	40
Tween 80	0.1	41
	1.0	31

The assays were done as described in fig. 4, except that detergents were added.
Triton: polyoxyethylene octylphenol;
Tween: polyoxyethylene sorbitan
Tween 20 (monolaurate), Tween 40 (monopalmitate), Tween 80 (monooleate).

It is noteworthy that this detergent does not disrupt the microsomal membrane. On the other hand, Triton X-100 strongly inhibits the enzyme and it disrupts the membrane. Three types of Tweens, also nonionogenic detergents, show intermediate effects.

When the microsomal suspension was treated with phospholipase A, the activity of the enzyme dropped with increasing solubilization of the membrane, as measured by the decrease of the turbidity (fig. 11).

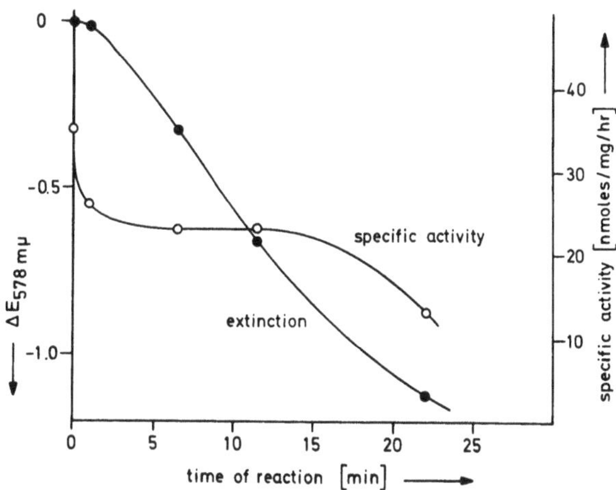

Fig. 11. Effect of treatment of microsomes with phospholipase A. The incubation with phospholipase A was performed at 25° C in a glas cell in a photometer Eppendorf. The cell contained in a total volume of 1.160 ml (pH 8.0): microsomal protein 10.4 mg/ml; cysteine 10 mM, serum albumin 5 mg/ml; $CaCl_2$ 10 mM. At t = 0 40 μl of a solution of 1 mg phospholipase A (from Crotalus terr. terr.) in 1 ml of 50 % glycerol (Boehringer,Mannheim) was added and the change in optical density at 578 mμ was measured against a blank which contained the same substances as the reaction mixture except phospholipase A. At different times after addition of phospholipase A samples of 10 μl were taken from the cell and assayed for HMG-CoA reductase activity (see legend with fig. 4).

From these experiments one gets the impression that under the conditions employed in these experiments the loss of the integrity of the membrane is always accompanied by loss of enzyme activity.

As mentioned above, the bile acids, when fed in the diet, cause a strong depression of both cholesterol synthesis (13) and HMG-CoA reductase activity (11). On the other hand, their in vivo effect cannot be explained by the effect found in vitro. Rather, the bile acids seem to interfere with the synthesis and/or the degradation of the HMG-CoA reductase protein.

ACKNOWLEDGMENT

We are greatly indebted to Mr. G. Greull for critically reading the manuscript.

REFERENCES

1. Knappe, J., Ringelmann, E. and Lynen, F., Biochem. Z. 332:195 (1959).
2. Kirtley, M.E. and Rudney, H., Biochemistry 6:230 (1967).
3. Hamprecht, B., Thesis. Universität München (1968).
4. Hamprecht, B. and Lynen, F., Europ. J. Biochem., 14:323 (1970).
5. Karmen, A., McCaffrey, J. and Bowman, R.L., J. Lipid Res. 3:372 (1962).
6. Regen, D.M., Riepertinger, C., Hamprecht, B. and Lynen, F., Biochem. Z. 346: 78 (1966).
7. Hamprecht, B., Naturwissenschaften 56:398 (1969).
8. Day, A.J., Fidge, N.H. and Wilkinson, G.N., J. Lipid Res. 7:132 (1966).
9. Pangborn, M.Y., J. Biol. Chem. 188: 471 (1951).
10. Bruckdorfer, K.R., Edwards, P.A. and Green, C., Europ. J. Biochem. 4: 506 (1968).
11. Hamprecht, B., Nüßler, C., Waltinger, G. and Lynen, F., in preparation.
12. Fimognari, G.M., Thesis. University of California (1964).
13. Back, P., Hamprecht, B. and Lynen, F., Arch. Biochem. Biophys. 133:11 (1969).
14. Miller, W.L. and Gaylor, J.L., Biochim. Biophys. Acta 137:400 (1967).

PROPERTIES AND LOCATION OF THE GTP-DEPENDENT ACYL-COENZYME A SYNTHETASE

Rossi C.R. and Carignani G.

from the Institute of Biological Chemistry

University of Padova, Padova (Italy)

Two different enzymatic systems are known for fatty acids activation in mitochondria.

In the "classical" systems ATP is the driving force (1-4) and inorganic pyrophosphate is the product generated by the hydrolysis of ATP as shown in the general equation 1):

1) R-COOH + CoASH + ATP ——> R-COSCoA + AMP + PP

More recently a second system has been reported (5-8). In this case the activation proceeds by using GTP as the energy donor and inorganic phosphate is the product of the hydrolysis of the nucleotide, as shown in equation 2):

2) R-COOH + CoASH + GTP ——> R-COSCoA + GDP + Pi

In the present paper the properties of the GTP-dependent acyl-CoA synthetase are described. In addition, in order to obtain some informations on its location within the mitochondrion, "structural" studies in submitochondrial fractions and "operational" studies in intact mitochondrial systems capable of oxidize fatty acids were undertaken. From both these two different approaches conclusive evidence has been obtained that the GTP-dependent acyl-CoA synthetase is sequestred in the "inner membrane-matrix" fraction, and that two different enzymes are present.
The first is specific for long-chain fatty acids and is

located in the inner membrane; another enzyme which is specific for short-chain fatty acids is very likely located in the matrix space.

PROPERTIES OF THE GTP-DEPENDENT ACYL-CoA SYNTHETASE

The GTP-dependent acyl-CoA synthetase was isolated and purified from liver mitochondria of Wistar strain albino rats (9). The isolation in soluble form was obtained by suspending the lyophilized mitochondrial pellets in 0.1 M KCl, 0.5 % triton X-100, $5x10^{-3}$M mercaptoethanol (9). The suspension was sonified (9) and the supernatant obtained after centrifugation was defined as Fraction I (Table I). The purification procedure (9) involved the adjustement to pH 3.4 and the readjustement to pH 7.0 without delay. The clear supernatant was defined as Fraction II (Table I). Further purification was obtained by removing inactive proteins by adsorption on calcium phosphate gel followed by precipitation with ammonium sulphate (9). The enzyme present in the supernatant was defined as Fraction III (Table I).

When lyophilized and extracted three times with 10 ml aliquots of 90 % acetone in water, the specific activity of Fraction II is uneffected in the presence of octanoate (see Table II) and reduced to a 21 % of the original specific activity with oleate as the sub-

TABLE I

GTP-dependent acyl-CoA synthetase. Specific activities of fractions prepared from rat liver mitochondria.

The values are expressed in nmoles/min./mg of protein. In the GTP-oleate system the increase in acyl-CoA concentration was measured (9),in the GTP-octanoate system the disappearance of sulphydryl groups was measured (9). The test system has been reported in a previous paper (9).

Enzymic system	Fraction		
	I	II	III
GTP-Oleate	1.2	5.0	34.0
GTP-Octanoate	5.8	22.4	67.8

strate (Table II). Preincubation of the acetone extracted protein (400 μg) with dipalmitoylphosphatidylcholine (40 μg) at 0° for 30 min. restored almost completely the activity with oleate (Table II). Similar results were obtained by preincubation of the enzymic protein with phosphatidylcholine previously removed by the acetone treatment (10). Phosphatidylethanolamine, phosphatidylserine, phosphatidic acid and sphingomyelyn had no effect. A smaller per cent of the initial activity was restored by increasing the temperature of preincubation (10). An aspecific effect of the temperature on the activity of the enzyme can be excluded because the purified enzyme is stable up to 55°.

TABLE II

GTP-dependent acyl-CoA synthetase. Effect of dipalmitoylphosphatidylcholine on the activity of acetone extracted enzyme.

Specific activities are expressed in nmoles/min./ /mg of protein. The test system and technical details as in Table I.

Sample	Additions	S.A. GTP Oleate	S.A. GTP Octanoate
Fraction II untreated	None	6.30	20.08
Fraction II acetone treated	None	1.27	21.00
Fraction II acetone treated	Dipalmitoyl-P-choline	5.02	19.67

On the assumption that the enzymic activity restored in the acetone treated enzyme by adding phosphatidylcholine is a function of the amount of protein which is saturated with phosphatidylcholine, termodynamic calculations on the binding can be made. The Arrhenius law was satisfied (11). By resolving the Van't Hoff relation (11) a change in enthalpy (negative value) of about -3 Kcal/mole and an entropy change (negative value) of about -12 entropy units were found (11).

These data indicate that the binding of phosphatidylcholine to the protein has the features of an esoergonic reaction giving a lipo-protein complex. The low negative entropy indicates that the complex has a lower number of degree of freedom which in turn suggests the possibility of electrostatic bonds between the polar groups of phosphatidylcholine and those of the protein molecule. It is possible that phosphatidylcholine, linked to the protein by electrostatic bonds, may confer chain-lenght specificity through its hydrophobic residues. In other words, long-chain fatty acids (in contrast with short-chain fatty acids) require the hydrophobic residues of phosphatidylcholine to facilitate their binding to the enzyme, possibly through conformational adaptivity of the active sites of the enzyme (11)

Fraction II was purified in a Sephadex G-75 column (9). From the column two components were isolated. The first was a protein (molecular weight 2×10^4), the second was a low molecular weight cofactor (9). When GTP-dependent acyl-CoA synthetase was assayed alternatively with the first or the second component alone no activity was detected (Table III). However, high activity with GTP and octanoate or oleate was obtained when the test system contained both the protein and the cofactor in the 1 to 1 ratio (Table III).

TABLE III

GTP-dependent acyl-CoA synthetase. Recombination of apoenzyme and cofactor.

Specific activities are expressed in nmoles/min./mg of protein. The test system and technical details as in Table I and II.

Additions	S.A.	
	GTP Oleate	GTP Octanoate
Protein	0	0
Cofactor from G-75 Sephadex	0	0
Protein + Cofactor from G-75 Sephad.	75	106
Protein + Cofactor from Biogel P_2	68	110
Protein + 4'-phosphopantotheine	86	102

At this stage of purification the cofactor contained sulfhydryl groups, organic phosphate and pantothenate (9), and gave the biuret reaction (9).

The cofactor isolated from Sephadex G-75 column was purified in a column of Biogel P_2 (9). Four different components were eluted: the first was inactive and positive in the biuret reaction. The second was not a polipeptide and it reactivated the apoenzyme isolated on a Sephadex G-75 column (Table III). This purified cofactor contained sulfhydryl groups, organic phosphate and pantothenate in a 1:1:1 ratio, as one could expect for 4'-phosphopantotheine.

The identification of the cofactor as 4'-phosphopantotheine is based on the following observations :
a) cofactor and 4'-phosphopantotheine show identical spectra with an adsorption maximum at 214 mμ and the same extinction coefficient (9);
b) cofactor and 4'-phosphopantotheine behave identically as far as the restoration of the activity of the apoenzyme is concerned.

4'-phosphopantotheine is presumably bound to the apoenzyme be weak secondary bonds since gel filtration alone effected the separation of the cofactor from the protein (9).

LOCATION OF GTP-DEPENDENT ACYL-CoA SYNTHETASE: OPERATIONAL STUDIES WITH INTACT MITOCHONDRIA

Mitochondria incubated in the absence of phosphate and in the presence of DNP efficiently oxidize added fatty acids (6,12,13) (Fig.1) although the ATP synthesis via electron transport chain was precluded by this inhibitor of oxidative phosphorylation. Under these conditions the GTP-dependent acyl-CoA synthesis appeared to be involved prior to the oxidation (12,13). Ortophosphate, a specific inhibitor of the GTP-dependent acyl-CoA synthetase (6), inhibited the oxidation of added fatty acids (Fig.1). This oxidation was also blocked by arsenate which is known to split succinyl--CoA and thus to prevent GTP formation via substrate level phosphorylation.

These data supported the view that acyl-CoA synthesis is tightly coupled to the oxidation of α-oxoglutarate and to succinyl-CoA synthesis according to the scheme shown in Fig. 1.

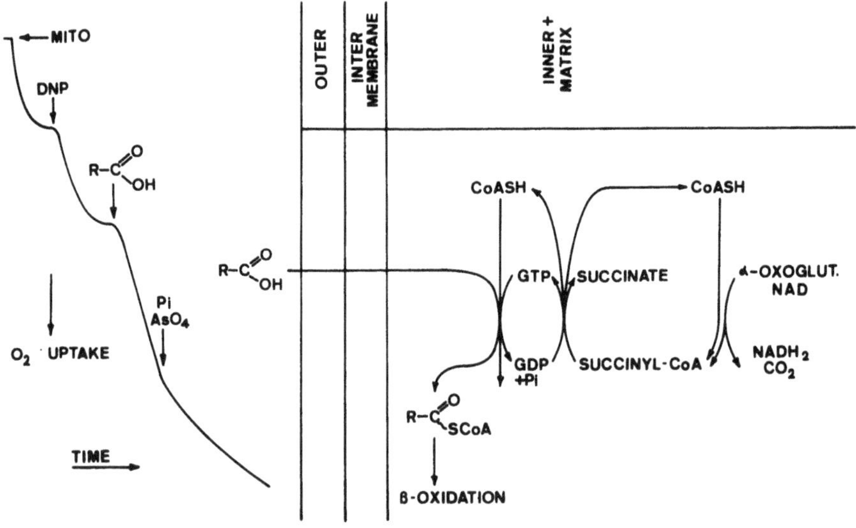

Figure 1 - Oxidation of fatty acids by intact rat liver mitochondria. On the left of the figure the oxygen uptake is shown. The incubation system contained in a volume of 2 ml: 10 mM Tris-buffer (pH 7.2), 26 mM NaCl, 58 mM KCl, 6 mM $MgCl_2$. At the points indicated by arrows, 6 mg of rat liver mitochondrial protein (MITO), 0.2 μmole dinitrophenol (DNP), 0.2 μmole oleate (R-COOH) 20 mM ortophosphate (PI) or 10 mM arsenate (ASO_4) were added. Oleate can be substituted with any other long- or short-chain fatty acid: different kind of fatty acid can induce different rate of oxygen uptake. In DNP-treated mitochondria no ATP can be detected. On the right of the figure a scheme of reactions presumably involved in the oxidation of fatty acid is illustrated. In this scheme the fatty acid oxidation, as well as the GTP synthesis via substrate level phosphorylation, is located into the "inner membrane - matrix" space.

In the oxidation of fatty acids driven by the GTP-dependent system no added CoASH, as well as no added GTP, were present in the system. "Endogenous" CoASH pool and "endogenous" GTP are sequestred into the"inner membrane-matrix" space (7,14). Within this compartment α-oxoglutarate oxidase and succinyl-CoASH synthetase systems are also located (7,15,16). On the other hand, the inner membrane is known to be impermeable to CoASH as well to acyl-CoA (7) and to GTP (14) so that fatty acids in order to be activated and oxidized, has to have direct accessibility to the "inner membrane-matrix" space. Accordingly, acyl-CoA formed via GTP-dependent system does not require the carnitine-linked transport mechanism (Fig. 1).

In conclusion, from the data obtained with operational studies a location of the GTP-dependent acyl-CoA synthetase into the "inner membrane-matrix" space can be anticipated.

LOCATION OF GTP-DEPENDENT ACYL-CoA SYNTHETASE: STRUCTURAL STUDIES

The location of GTP-dependent acyl-CoA synthetase in the context of the mitochondrial structure has been studied by following the enzyme concentration in submitochondrial fractions obtained by digitonin treatment as described by Schnaitman and Greenawalt (17).

By treating rat liver mitochondria with low concentration of digitonin the outer membrane is stripped (19). The "low speed pellet" obtained by centrifuging at 9,000 x g is generally considered to consist of intact inner membrane plus matrix (17). This fraction should be free of any significant amount of outer membrane or intact mitochondria (17). In our hands (Table IV) about 97 per cent of the total activity of succinate dehydrogenase (a marker enzyme for the inner membrane (16-18))and 80 per cent of malate dehydrogenase (the marker enzyme for the matrix (16-19) were found in this fraction (Table IV). About 27 per cent of monoamino oxidase could represent contamination by outer membrane (16,17,20).

The "high speed pellet" obtained by centrifugation of the digitonin treated mitochondria at 144,000 x x g (17) contained a rather high per cent of monoamino oxidase (Table IV) with a very high specific activity, thus indicating that this fraction is predominantly formed of disrupted external mitochondrial membranes (17).

As the distribution of GTP-dependent acyl-CoA synthetase is concerned, Table IV shows that most of the activity (over 80 per cent of the total activity) with short as well as with long-chain fatty acids was concentrated in the "low speed pellet", i.e. in the frac-

TABLE IV

Distribution of mitochondrial enzymes after digitonin treatment

All specific activities (S.A.) are expressed as nmoles/min/mg of protein. Monoamino oxidase, succinate dehydrogenase, malate dehydrogenase were determined as described by Schnaitman and Greenawalt (17). The test system for the GTP-dependent acyl-CoA synthetase has been reported in a previous paper (9) : before the analysis each fraction was lyophilized, suspended in 0.1 M KCl, 0.5% Triton X-100, $5x10^{-3}$M CH_3CH_2SH and sonicated for 5 min. (20,000 kilocycles) at temperature between 0° and 5°.

SAMPLE	Monoamino oxidase		Succinate dehydrogenase		Malate dehydrogenase		GTP-dependent Acyl-CoA synthetase			
							GTP oleate		GTP octanoate	
	S.A.	%	S.A.	%	S.A.	%	S.A.	%	S.A.	%
Whole Mitoch.	25.3	100	645	100	2250	100	0.95	100	6.30	100
Low speed P	9.8	26	1000	97	5720	80	1.33	85	8.20	82
High speed P	212.0	48	200	1	1200	1	0.78	4	13.00	11
High speed S	20.0	27	0	0	2240	16	0.01	0	0.80	4
Recovery		101		98		97		89		97

Low speed P = pellet obtained by centrifuging at 9,000 x g
High speed P = " " " " " 144,000 x g
High speed S = supernatant obtained after centrifugation at 144,000 x g

tion where succinate and malate dehydrogenase are specifically located.

These results nicely agree with those obtained with operational studies and confirm that the GTP-dependent acyl-CoA synthetase is sequestred into the "inner membrane-matrix" space.

In order to obtain finer informations on the location of the GTP-dependent acyl-CoA synthetase, the "low-speed pellet" - i.e. the "inner membrane-matrix" fraction - was suspended in 0.1 M buffer phosphate (pH 7.2), and subjected to sonic treatment as described by Brdiczka et al. (21). This treatment opens up the inner mitochondrial membrane which is recovered in the 144,000 x g pellet (21). The supernatant contained the soluble enzymes of the matrix (21) and the enzymes which, loosely attached to the inner membrane, were released following the sonic treatment (17). Both the supernatant and the pellet were dialyzed overnight against 1 liter of 0.1 M KCl to remove the phosphate, which, as previously reported (6), inhibits the GTP-dependent acyl-CoA synthetase.

Data summarized in Table V show that malate dehydrogenase has been found predominantly (80 per cent of its total activity) in the supernatant. In this fraction has also been found some 80 per cent of the GTP--dependent acyl-CoA synthetase in the presence of short-chain fatty acids as substrates. With long-chain fatty acids no activity could be detected not even in the presence of added phosphatidylcholine.

With long-chain fatty acids the activity is concentrated (over 99 per cent of the total activity) in the 144,00 x g pellet. This pellet contains mostly fragments of inner membrane as demonstrated by the concentration in this fraction of the succinate dehydrogenase (Table V).

It has been previously reported that the GTP-dependent acyl-CoA synthetase solubilized and purified from intact mitochondria is active both with short and with long-chain fatty acids. With long-chain fatty acids bound phosphatidylcholine was needed probably by facilitating the binding of fatty acid to the enzyme. Data obtained from submitochondrial fractions demonstrate that two different enzymes do exist: one, specific for long-chain fatty acid and bound to phosphatidylcholine in a lipoprotein complex, is firmly

TABLE V

Distribution of mitochondrial enzymes in digitonin particles

All specific activities (S.A.) are expressed in nmoles/min/mg of protein. The test system and technical details as in Table IV.

SAMPLE	Succinate dehydrogenase		Malate dehydrogenase		GTP-dependent Acyl-CoA synthetase			
					GTP-Oleate		GTP-Octanoate	
	S.A.	%	S.A.	%	S.A.	%	S.A.	%
Whole digitonin particles	740	100	5900	100	2.0	100	5.1	100
Pellet	1675	75	2000	11	7.3	99	3.3	21
Supernatant	250	22	7000	79	0	0	5.4	70
Recovery		97		90		99		91

attached to the inner membrane, the other one, specific for short-chain fatty acids, is released in solution by sonication and, hence, probably located in the matrix or loosely attached to the inner membrane.

Our data for the enzyme specific with short-chain fatty acids are in agreement with those reported by Garland, Haddock and Yates (7). These Authors separated the two mitochondrial membranes by density gradient centrifugation following mitochondrial swelling and contraction (20). The GTP-dependent acyl-CoA synthetase working with decatetraenoic acid was found in the soluble fraction.

As the enzyme specific with long-chain fatty acids is concerned our data are in contrast with those reported by Allman et al. (8). In these experiments the outer and the inner membranes were separated by a procedure involving a disruption of the mitochondria with phospholipase. By this procedure the GTP-dependent acyl-CoA synthetase specific with long-chain fatty acids has been found concentrated in the outer membrane. However, in the case of phospholipase treatment a redistribution of some enzymes cannot be excluded (18). In other words, phospholipase could "relocate" some enzymes from the inner to the outer membrane. A similar finding has been reported for the β-hydroxybutyrate dehydrogenase (22), an enzyme which, as the GTP-dependent synthetase specific for long-chain fatty acids, is linked to a phospholipid (23). Such an enzyme has been recovered in the outer membrane fraction following the phospholipase treatment and in the inner membrane fraction after sonication of mitochondria (22).

ACKNOWLEDGEMENTS

The Authors express their gratitude to Mrs. Maria Rosa Badomer Zuin for her skilled technical assistance.

This investigation was supported by C.N.R., Centro per lo Studio della Fisiologia dei Mitocondri.

REFERENCES

1. Jones, M.E., Black, S., Flynn, R.M. and Lipman, F., Biochim. Biophys. Acta 12:141 (1953).
2. Mahler, H.R., Wakil, S.T. and Bock, R.M., J. Biol. Chem. 204:453 (L953).
3. Kornberg, A. and Pricer, W.E., J. Biol. Chem.204:329 (1953).
4. Green, D.E. and Gibson, D.M., in "Metabolic Pathway" (D.M. Greenberg, ed.), Vol. I, 2nd ed., Academic Press, p. 301 (1969).
5. Rossi, C.R. and Gibson, D.M., J. Biol. Chem. 239: :1694 (1964).
6. Galzigna, L., Rossi, C.R., Sartorelli, L. and Gibson, D.M., J. Biol. Chem. 242:2111 (1967).
7. Garland, P.B., Haddock, B.A. and Yates, D.W., in "5th Meeting Federation of European Biochemical Societies" (L. Ernster and Z. Drahota, eds.), p.111, Prague, July 1968, Academic Press, London and New York (1969).
8. Allmann, D.W., Galzigna, L., McCaman, R.E. and Green, D.E., Arch. Biochem. Biophys. 117:413 (1966).
9. Rossi, C.R., Alexandre, A., Galzigna, L., Sartorelli, L. and Gibson, D.M., J. Biol. Chem. in press.
10. Sartorelli, L., Galzigna, L., Rossi, C.R. and Gibson, D.M., Biochem. Biophys. Res. Comm. 26:90 (1967)
11. Galzigna, L., Sartorelli, L., Rossi, C.R., and Gibson, D.M., Lipids 4:459 (1969).
12. Van den Bergh, S.G., Biochim. Biophys. Acta 98:442 (1965).
13. Rossi, C.R., Alexandre, A., Sartorelli, L., European J. Biochem. 4:31 (1968).
14. Klingenberg, M. and Pfaff, E., in "Regulation of Metabolic Processes in Mitochondria" (J.M. Tager, S. Papa, E. Quagliariello and E.C. Slater,eds.), p. 180, Elsevier Publ. Co., New York (1966).
15. Nichols, D.G., Shepherd, D. and Garland, P.B., Biochem.J. 103:677 (1967).
16. Ernster, L. and Kuylensterna, B., in "Proc. 5th Meeting Federation of European Biochemical Societies" (L. Ernster and Z. Drahota, eds.), Praghe, July 1968, p. 5, Academic Press, London and New York, (1969).
17. Schnaitman, C. and Greenawalt, J.W., J. Cell Biol. 38:158 (1968).
18. Parsons, D.F., Williams, G.R. and Chance, B., Ann. New York Acad. Sci. 137:643 (1966).
19. Green, D.E., Discussion in "Mitochondrial Structure and Compartmentation" (E. Quagliariello, S. Papa, E.C. Slater and J.M. Tager, eds.), p.118, Adriatica

Editrice, Bari (1967).
20. Sottocasa, G.L., Kuylenstierna, B., Ernster, L. and Bergstrand, A., in "Methods in Enzymology" (R.W. Estabrook and M.E. Pullman, eds.), Vol. X, p. 448, Academic Press, London and New York (1967).
21. Brdiczka, D., Gerbitz, K. and Pette, D., European J. Biochem., 11:234 (L969).
22. Green, D.E., Bachmann, E., Almann, D.W. and Perdue, J.F., Arch. Biochem. Biophys. 115:172 (1966).
23. Sekuzu, I., Jurtshuk, P. and Green, D.E., J. Biol. Chem. 238:975 (1963).

ON THE ROLE OF IRON IN THE LINKAGE OF SUCCINATE DEHYDROGENASE TO THE MEMBRANE[a,b]

PAOLO CERLETTI and GIULIANA ZANETTI

Istituto di Biochimica Generale, Università di Milano

Milano (Italy)

Little is known about the nature of the linkage between succinate dehydrogenase (succinate:(acceptor) oxidoreductase, EC 1.3.99.1) and the inner mitochondrial membrane. A role of iron has been proposed on the basis of the solubilizing effect of cyanide and of hydroxyl ions on the flavoprotein (1). This action has been interpreted as due to replacement of a ligand in a non-heme coordination complex (1). Direct evidence for this interaction has so far been lacking but some relevant results on the point were recently obtained in our laboratory.

Information on the binding can also be gained by dissociating the flavoprotein from the membrane using other agents or else in reconstitution studies in which the purified flavoprotein recombines with other isolated components of the membrane to produce a functioning integrated complex. Data on the state of iron in the isolated flavoprotein are also important.

We shall discuss our results in the light of information obtained by other approaches.

---

a This work has been supported in part by grants from the Italian National Research Council.

b Abbreviations : TTA, 2-thenoyltrifluoroacetone,4,4,4-trifluoro-1-(2-thienyl)1-3-butane-dione ; DCPI, 2,6-dichlorophenolindophenol.

Dissociation of Succinate Dehydrogenase from the Membrane

The difficulty to dissociate the dehydrogenase from particulate preparations and its high affinity for the other reactants in reconstitution experiments (1,3) indicate the existence of tight connections between succinate dehydrogenase and other membrane constituents. The recombination is extremely rapid after a lag of 40-100 sec (ref. 4).

Organic solvents (5,6), alkali (5), and cyanide (5), resolve the membrane system containing the flavoprotein. Soluble preparations of the flavoprotein obtained by different procedures are satisfactorily uniform in composition, catalytic properties and in proper conditions, reconstitutive capacity. In these soluble preparations, the dehydrogenase behaves in many respects like it does in the membrane. Indeed a) similar intermediates are probably involved in the redox process (7,8)(this point is still debated (9,10) b) the ratio flavin:iron:labile sulfide (1:8:8) is the same in the soluble flavoprotein (2,11) and in succinate-ubiquionone reductase (12), the smallest fragment of the respiratory assembly containing the flavoprotein c) both forms of enzymes, soluble and particulate, undergo the so called "activation", i.e. they increase their catalytic efficiency when preincubated with succinate, malonate, oxaloacetate, phosphate (10,13).

Some properties, however, are modified, namely:

1 the catalytic centre activities of the membrane-bound and the soluble dehydrogenase differ (14).

2 the iron chelator TTA below millimolar concentrations inhibits only the membrane-bound enzyme (15).

3 the membrane-bound dehydrogenase is more stable (2,5,16).

4 the ability to react with DCPI is lost during solubilization (2).

5 a further difference is the so called "thermal activation". Heating in the absence of the above mentioned effectors increases the activity of the membrane-bound dehydrogenase while this is not the case for the soluble one (17). Preliminary results from our laboratory suggest that this phenomenon may be related to changes in permeability of the membrane (18).

General preparative damage during the purification of the soluble flavoprotein plays a minor role in these differences. They are specifically related to the separation of the flavoprotein from its native hydrophobic environment, i.e. they depend upon the interaction of the flavoprotein with other membrane components (14, 19).

Some of the treatments which break down the membrane irreversibly modify the particle, and the flavoprotein alone is fit for reconstitution. Other dissociating procedures separate also the rest of the system in conditions to be tested for recombination.

Effect of solvents. The solubilizing agents most commonly used are acetone (6) and butanol (5). In either case the effect does not depend upon the pH (neutral or alkaline) at which the treatment is made (19). Most lipid of the particulate preparation is extracted and removed with the solvent. Some is left in the soluble preparation (14) but it is removed as the purification proceeds (20).

The action of the solvent profoundly involves all the hydrophobic components of the membrane and the system interacting with the flavoprotein in the membrane is irreversibly destroyed.

Dissociation by alkali. An appreciable dissociation of the dehydrogenase from the membrane occurs above pH 9.4 (ref.5). The treatment is usually carried out between pH 9.5 and 10.0 to avoid inactivation of the enzyme. The particle from which the flavoprotein has dissociated can be isolated and used for reconstitution experiments (5). No direct evidence is however available about the interaction of hydroxyl ions with iron.

Incorporation of Cyanide

Cyanide dissociates succinate dehydrogenase from the membrane independently of the pH at which treatment is made. The action depends upon temperature and is rather slow (21). Succinate dehydrogenase activity decreases during treatment (22):this is due to the different catalytic centre activity between membrane-bound and soluble succinate dehydrogenase (14) and perhaps also to thermal inactivation during preincubation with cyanide. The solubilized flavoprotein is inactive in reconstitution (5) and its catalytic centre activity is approximately the same as for other reconstitutively inactive forms of succinate dehydrogenase (23). The cyanide enzyme is activated by succinate.

The dissociated flavoprotein contains about 6 moles cyanide per mole flavin.[a] As shown in Table I, the binding is remarkably stable and it is maintained throughout the purification procedures (ammonium sulfate fractionation, gel filtration, electrofocusing) (23). Cyanide reacts as well with the flavoprotein solubilized by solvents or by alkali : the same amount is incorporated as during dissociation of the membrane-bound flavoprotein by cyanide. The soluble flavoprotein treated with cyanide looses its reconstitutive capacity and the catalytic activity decreases (23).

The results so far reported do not provide direct evidence for the interaction of cyanide with iron. To this purpose we applied to the cyanide-solubilized enzyme various treatments which release iron from the flavoprotein : cyanide is also released and, as shown in Table II, cyanide and iron residual in the flavoprotein after each treatment are in good stoichiometry (23). These data favour the idea that cyanide interacts with non-heme iron in the flavoprotein.

During cyanide treatment of the membrane, cyanide is bound also to the particle from which flavoprotein is dissociated. About 1 mole $^{14}CN^-$ per g protein is incorporated rapidly (approximately 8 moles/mole original peptide-bound flavin), and up to 6 times more cyanide is then slowly bound (24). The incorporation of cyanide in this case is aspecific with respect to the interactions with the flavoprotein,and the cyanide-particle is fit for reconstitution (1,5). A similar situation is observed with iron chelators like L-histidine, o-phenanthroline and TTA. They prevent reincorporation of the soluble flavoprotein into succinate dehydrogenase depleted particles but do not affect the reconstitutive capacity of the particles (25). All these results show that iron chelation modifies the binding properties of the flavoprotein, while those of the particles are not affected.

If the suggestion is accepted that hydroxyl ions and cyanide replace a ligand on non-heme iron in the flavoprotein, an important question arises as to why the effects on the flavoprotein are so different, the one solubilized by alkali being active in reconsti-

(a) Lee and King (24) reported incorporation of 5 moles cyanide per mole flavin. These authors,however,did not quantify the flavin, but deduced the amount of enzyme from activity measurements.

TABLE I

CYANIDE CONTENT OF CYANIDE–SOLUBILIZED SUCCINATE DEHYDROGENASE DURING PURIFICATION

The washed Keilin Hartree heart muscle preparation was incubated 45 min at 30° with 50 KCN in 50 mM phosphate–50 mM borate buffer, pH 8. Assayes are described in ref.14. Succinate dehydrogenase activity was measured after activation (14). It is given in micromoles succinate oxidized/min at 25° and 2 mM phenazine.

	total protein mg	activity in micromoles/mg	peptide bound flavin (PBF) nmoles/mg protein	bound CN^- moles/mole PBF
Keilin Hartree (K.H.) preparation	13,300	0.9		
Washed K.H. prep.	4,470	1.5		
Supernatant after CN^- treatment	343	3.5		
Precipitate 30–65% satn. of $(NH_4)_2SO_4$	187	4.8	1	8–10
Effluent from Sephadex G 200	60	9.2	2	6–7
Electrofocusing				
fraction I.P. pH 5.7	10	4.0	2.5	6–7
fraction I.P. pH 6.1	10	3.5	2.8	5

TABLE II

CYANIDE AND IRON RELEASE UNDER DIFFERENT CONDITIONS

Purified succinate dehydrogenase, solubilized by cyanide, was submitted to various procedures used to release non-heme or total iron from proteins. Iron was determined according to Van de Bogart and Beinert (31) in the protein before and after treatment and in the supernatant derived from treatment. Cyanide was counted in the same fractions.

Treatment applied	Ref.	CN^-			Fe	
		before treatment	after treatment released	after treatment bound	after treatment bound	before treatment
10% TCA, 15 min 0°	32	6.5	1.7	4.9	4.2	8.4
Extraction with 50% acetic and 1.25% mercapto-acetic acid at 20°	33	5.4	1.7	3.7	2.2	8.4
1 N HCl, 1 hour 100°	34	5.4		0–0.3	0–0.5	8.4
0.3 N HCl, 10 min 80°	35	5.4		1.4	1.6	8.4

tution (5), the one treated with cyanide being irreversibly modified (5). Probably cyanide binds iron much more firmly and it is not displaced in the conditions in which reconstitution is performed.

Modifications Involving Iron and the Flavoprotein

There are data showing that modifications in the flavoprotein may affect the reactivity with cyanide.

When the system is reduced by succinate or by other agents like NADH and dithionite, cyanide does not dissociate the flavoprotein from the membrane (1,21). Cyanide nevertheless is incorporated into the particle (24): the flavoprotein can be solubilized from it by solvent extraction and it is found to contain only a small amount of cyanide (23). Also the incorporation into the soluble flavoprotein is decreased in the presence of succinate (23); the data are summarized in table III. The reaction of cyanide with iron seems therefore impaired in the reduced flavoprotein. This might be due to a different conformation of the reduced flavoprotein.

The linkage of other respiratory carriers to the membrane is affected by their redox state. For instance reduced NADH dehydrogenase is not dissociated from complex I by chaotropic agents (26). The case is not as clear with succinate dehydrogenase: indeed hydroxyl ions and chaotropic agents (27) resolve the reduced system but the presence of succinate inhibits the extraction of the flavoprotein from complex II by freezing and thawing at pH 8 (ref 10). Reducing conditions at the moment of disconnecting the flavoprotein from the membrane, however, affect specifically the reconstitutive properties of the solubilized dehydrogenase. Solvents or alkali under non-reducing conditions yield a modified soluble preparation which does not reconstitute electron transport to other physiological carriers (5). In subsequent steps of purification (treatment with calcium phosphate gel, ammonium sulphate fractionaction) succinate can be removed without losing the reconstitutive capacity (28). On the other hand the enzyme disconnected from the membrane in the absence of succinate if subsequently treated with succinate undergoes activation but remains reconstitutively inactive (15). Activation and loss of reconstitutive capacity represent different modifications of the flavoprotein: indeed beside succinate, non-reducing effectors

TABLE III

EFFECT OF SUCCINATE ON THE INCORPORATION OF CYANIDE

Heart muscle preparations and purified succinate dehydrogenase (SDH) were incubated anaerobically for 45 min at 30° with 50 mM KCN containing 3–20 µC/ml $K^{14}CN$ with the additions indicated. When cyanide acted on the particle in the presence of succinate, butanol treatment was subsequently applied to release the dehydrogenase which was purified as described in ref. 20. Cyanide was measured in the soluble purified flavoprotein. Results are given as moles cyanide incorporated per mole peptide-bound flavin in the preparation.

preparation treated with cyanide	cyanide incorporated additions during cyanide treatment	
	none	succinate 40 mM
soluble succinate dehydrogenase	5–6	2.5
heart muscle preparation: analyzed		
SDH solubilized by cyanide	6	no solubilization
SDH solubilized by butanol		1.2
cyanide particle[a]	6.5	6.9

(a) calculated from data of Lee and King (24).

like fumarate, malonate and phosphate produce activation (13), and it occurs also in reconstitutively inactive preparations without making them fit to recombine (5).

According to WANG (25) the flavoprotein which reacted with cyanide, binds to the membrane but does not produce a functioning system. Reconstitutively inactive preparation behave similarly: they interact with cytochrome b but do not reconstitute electron transport to ubiquinone (15). This might indicate that in either case the properties of the flavoprotein as concerns general interactions with the membrane are not drastically changed. The flavoprotein is, however, modified in a way that prevents the reconstitution of function.

It is difficult to equate the fitness of the flavoprotein to reconstitution with any of the known parameters of the iron-sulfur system. Reconstitutive capacity decays more rapidly than catalytic centre activity towards artificial acceptors and ESR signals at g =2.01 and g = 1.94 (ref. 5 and 29). Flavin, iron and labile sulfide are in the same ratio 1:8:8 in the purified, reconstitutively inactive enzyme (3), when carefully prepared, as in the reconstitutively active one (11). Baginsky and Hatefi (3) have shown that the reconstitutive fitness is restored to the soluble reconstitutively inactive flavoprotein by treating it with ferrous ammonium sulfate, sodium sulfide and mercaptoethanol. The difference in reconstitutive capacity resides, therefore, in a more subtle modification of the flavoprotein which is counteracted by treating it with iron and sulfide. It is possible that incorporated colloidal iron-sulfide complex may bridge the flavoprotein structurally and functionally to the particle.

Cyanide and Other Iron Chelators

The action of cyanide differs in some respects from that of other iron chelators. Polidentate chelators do not dissociate either reduced or oxidized flavoprotein from the particle (25). Apparently the iron involved in binding is not available to a polidentate chelator in the particle. Once the linkage to the membrane is resolved, the iron is chelated regardless of the redox state of the flavoprotein and this prevents recombination with the membrane (25).

Differences are also found in the action of these agents on the catalytic activity of the enzyme. Cyanide does not inhibit either soluble or particulate succinate dehydrogenase. TTA and other lipophilic iron chelators, at millimolar concentrations, inhibit both forms of the enzyme, but below 100 micromolar, inhibit only the membrane-bound dehydrogenase (15) : this is because of the lack of a hydrophobic environment in purified soluble preparations (14,30). Indeed, addition of phospholipids allows inhibition by TTA of soluble preparations as well (14).

Interactions among Membrane Constituents

Numerous facts suggest, therefore, a role of iron in the linkage to the membrane. The flavoprotein seems, however, to participate also in the binding through some other relevant property, probably depending upon its conformational state. There are no data which can indicate a direct interaction of the iron with other constituents of the membrane system. Unfortunately, these membrane participants in the linkage are less well characterized than the flavoprotein. However, some information on the type of interaction occurring in the integrated system is available.

The reactivity of cytochrome b, factor F_4 (a protein which is supposed to participate in the organization of the enzyme involved in oxidative phosphorylation) and of phospholipids with the flavoprotein has been tested (2,14). Each of them binds purified succinate dehydrogenase and this restores, to a varying extent, the properties modified during disconnection from the membrane. The interactions are specific for the flavoprotein : enzymes which are not mitochondrial are not bound. Further, the flavoprotein and cytochrome b reciprocally modify each other in the interaction (2). However, all the mentioned components, together are required to restore reactivity with DCPI and ubiquinone though neither cytochrome b nor F_4 seems to participate in the electron transfer (2). Some characteristics required in the interacting compounds are established : not all phospholipids bind to the flavoprotein (14) and only crude mitochondrial phospholipids or purified lecithin or phosphatidylethanolamine, but not cardiolipin, restore succinate-ubiquinone reductase activity (2). Protein structure in cytochrome b is important for binding succinate dehydrogenase and a decreased heme content or modifications introduced by detergents prevent to reconstitute the transfer to ubiquinone.

Lipids alone are not the major factor for binding to the membrane: indeed more than 90% hydrolysis of phospholipids by phospholipase A or C does not dissociate the dehydrogenase from the membrane (30). They are probably important for the stability of the interacting macromolecules (2).

Multiple interactions appear therefore to affect flavoprotein integration within the membrane. Binding to the membrane does not imply reconstitution of function; the latter probably indicates a higher level of structural integrity of the system. Likely, general protein-protein and protein-lipid interactions are the main factors in structuring the system and changes in the iron coordination sphere affect the conformation of the native flavoprotein and thereby influence the binding to the membrane.

SUMMARY

The incorporation of cyanide into soluble succinate dehydrogenase and into the flavoprotein solubilized by cyanide treatment was studied. The amount incorporated depends on the redox state of the enzyme.

The cyanide-solubilized dehydrogenase was purified: a constant ratio of bound cyanide to flavin was found throughout purification. The purified flavoprotein was submitted to various procedures to release the iron: only partial extraction was obtained. Cyanide was also liberated in part. Residual bound cyanide and iron were stoichiometric to each other.

The data suggest that cyanide is bound to non-heme iron in the flavoprotein. The role of iron in the linkage of succinate dehydrogenase to the inner mitochondrial membrane is discussed: it is suggested that the metal does not participate directly to the linkage.

REFERENCES

1. King, T.E., "Advances in Enzymology" (F.F. Nord ed.), vol.28, p. 155, Interscience, New York (1966).
2. Bruni, A. and Racker, E., J.Biol.Chem. 243: 962 (1968)
3. Baginsky, M.L. and Hatefi, Y., J.Biol.Chem. 244:5313 (1969).
4. Lee, L.P., Estabrook, R.W. and Chance, B., Biochim.Biophys. Acta, 99: 32 (1965).
5. King, T.E., J.Biol.Chem. 238; 4037 (1963).
6. Bernath, P. and Singer, T.P., in "Methods in Enzymology" (S.P. Colowick and N.O. Kaplan eds.), vol.5, p. 597, Academic Press, New York (1962).
7. Beinert, H. and Sands, R.M., Biochem.Biophys.Res.Commun. 3:41 (1960).
8. Dervartanian, D.V., Veeger, C., Orme-Johonson, W.H. and Beinert, H., Biochim.Biophys.Acta 191:22 (1969).
9. Gawron, O., Mahajan, K.P., Limetti, M., Kananen G. and Glaid A.J. III, Biochemistry, 5: 4111 (1966).
10. Zeylemaker, W.P., Dervartanian, D.V., Veeger, C. and Slater, E.C., Biochim.Biophys.Acta 178: 213 (1969).
11. King, T.E., Biochem.Biophys. Res. Commun. 16: 41 (1964).
12. Ziegler, D.M. and Doeg, K.A., Arch.Biochem.Biophys. 97:41 (1962).
13. Kimura, T., Hauber, J. and Singer, T.P., J.Biol.Chem., 242: 4987 (1967).
14. Cerletti, P., Giovenco, M.A., Giordano, M.G., Giovenco, S. and Strom, R., Biochim.Biophys.Acta 146: 380 (1967).
15. Redfearm, E.R., Whittaker, P.A. and Burgos, J., in "Oxidases and Related Redox Compounds" (T.E. King, M.S. Mason and M. Morrison eds.), p. 943, Wiley, New York (1965).
16. Cerletti, P., Cajafa, P., Giordano, M.G. and Testolin, G., Lipids, in press.
17. Thorn, M.B., Biochem.J., 85: 116 (1962).
18. Cerletti, P. and Rossi, C., unpublished results.
19. Cerletti, P., Giovenco, S., Testolin, G., and Binotti, I., in "Membrane Models and the Formation of Biological Membranes" (L. Bolis and B.A. Pethica eds.), p. 166 North Holland Amsterdam (1968).
20. Cerletti, P., Zanetti, G., Testolin, G., Rossi, C., Rossi, F., and Osenga, G., in "Flavins and Flavoproteins" (H. Kamin ed.) University Park Press, in press.
21. Wu, G.T. and King, T.E., Federat. Proc. 26: 732 (1967).

22. Giuditta, A. and Singer, T.P., J.Biol.Chem. 234: 662 (1959).
23. Zanetti, G. and Cerletti, P., to be published.
24. Lee, C.P. and King, T.E., Biochim.Biophys.Acta 59: 716 (1962).
25. Wang, T.Y. and Wang, Y.L., Scientia Sinica (Peking) 13: 1799 (1964).
26. Davis, K.A. and Hatefi, Y., Biochemistry 8: 3355 (1969).
27. Hatefi, Y., Davis, K.A., Hanstein, W.G., and Ghalambor, M.A., Arch.Biochem.Biophys. 137: 286 (1970).
28. Veeger, C., Dervartanian, D.V. and Zeylemaker, W.P., in "Methods in Enzymology" (J.M. Lowenstein ed.), vol. 13, p.81, Academic Press, New York (1969).
29. King, T.E., Howard, R.L. and Mason, H.S., Biochem.Biophys.Res. Commun. 5: 329 (1961).
30. Cerletti, P., Cajafa, P., Giordano, M.G. and Giovenco, M.A., Biochim.Biophys.Acta 191, 502 (1969).
31. Van de Bogart, M. and Beinert, H., Analyt. Biochem. 20: 325 (1967).
32. Massey, V., J.Biol.Chem., 229: 763 (1957).
33. Doeg, K.A. and Ziegler, D.M., Arch.Biochem.Biophys. 97: 37 (1962).
34. King, T.E., Nickel, K.S. and Jensen, D.R., J.Biol.Chem., 239: 1989 (1964).
35. Lovenberg, W., Buchanan, B.B. and Rabinowitz, J.C., J.Biol. Chem., 238: 3899 (1963).

THE ROLE OF THE PHOSPHORYLATED INTERMEDIATE IN THE REACTION OF THE (Na^+ + K^+)-ACTIVATED ENZYME SYSTEM

J. C. Skou

Institute of Physiology, University of Aarhus

Denmark

Membrane fractions isolated from cells which exhibit a potassium coupled active transport of sodium contain a (Na^+ + K^+)-activated enzyme system. This system fulfils so many of the requirements for a sodium-potassium transport system that it is reasonable to assume that the system is involved in the active transport of the cations (see reviews by Skou[1]; Heinz[2]; Albers[3]; Glynn[4]).

The substrate for the system is ATP. It requires a combined effect of sodium and potassium for activation. The activating effect of the two ions is on different sites of the system. Sodium activates from a site which in the intact cell is located on the inside of the membrane while potassium activates from a site which is located on the outside of the membrane[5,6,7,8]. Potassium competes for sodium at the sodium site, and sodium for potassium at the potassium site[1]. The apparent affinity for sodium at the sodium site is about 3-4 times the affinity for potassium while the apparent affinity for potassium at the potassium site is about 60-100 times the affinity for sodium[9,10].

THE INTERMEDIARY STEPS IN THE REACTION

With magnesium and sodium, but no potassium, the reaction of the system with ATP leads to the formation of a phosphorylated intermediate; addition of potassium to a prephosphorylated system leads to a dephosphorylation (for ref. see Albers[3], Skou and Hilberg[11]).

$$E + ATP \rightleftharpoons E\text{-}ATP \quad (1)$$

$$E\text{-}ATP \xrightleftharpoons{Mg^{++}, Na^{+}} E{\sim}P + ADP \quad (2)$$

$$E{\sim}P + H_2O \xrightarrow{K^{+}} E + Pi \quad (3)$$

$$E + ATP + H_2O \longrightarrow E + ADP + Pi \quad (4)$$

From the observation that the requirement for magnesium for the sodium-dependent ATP-ADP exchange catalyzed by the system is much lower than for the ($Na^+ + K^+$)-dependent hydrolysis, Fahn et al.[12] suggested the following scheme for the reaction.

$$E + Mg^{++} \rightleftharpoons Mg\text{-}E \quad (5)$$

$$Mg\text{-}E + ATP \xrightleftharpoons{Na^{+}} Mg\text{-}E{\sim}P + ATP \quad (6)$$

$$Mg\text{-}E{\sim}P + Mg^{++} \rightleftharpoons Mg\text{-}E{\sim}P\text{-}Mg \quad (7)$$

$$Mg\text{-}E{\sim}P\text{-}Mg \xrightleftharpoons{(Na^{+}\ ?)} Mg\text{-}E\text{-}P\text{-}Mg \quad (8)$$

$$Mg\text{-}E\text{-}P\text{-}Mg + H_2O \xrightarrow{K^{+}} Mg\text{-}E + Pi + Mg^{++} \quad (9)$$

In this scheme the enzyme system reacts with two magnesium molecules, and the affinity for the first is an order of magnitude higher than for the second. The hydrolysis goes via formation of two phosphorylated intermediates and the magnesium molecule for which the system has the highest affinity is necessary for the reaction with ATP and for the formation of the first phosphorylated intermediate, Mg-E P. The second magnesium molecule is necessary for the transformation of the Mg-E P into the second phosphorylated intermediate, Mg-E-P-Mg. The two phosphorylated intermediates differ in that Mg-E P can react with ADP, while Mg-E-P-Mg cannot, but can react with potassium and be dephosphorylated.

Evidence for the existence of the two predicted phosphorylated intermediates has been given by Post et al.[13]. In experiments at 0° C they were able to show that the phosphorylated intermediates formed with a concentration of magnesium which was low and high respectively relative to the concentration of ATP, differed in their reactivity towards ADP, potassium and g-strophanthin. When formed with a low concentration of magnesium, the addition of ADP leads to an increased rate of dephosphorylation, while potassium has a low or no effect. When formed with a high magnesium concentration, the addition of potassium leads to an increased rate of dephosphorylation, while ADP has a low or no effect. The ADP sensitive phospho-enzyme does not react with g-strophanthin, while the potassium sensitive does.

According to the scheme by Fahn et al.[12] the enzyme system reacts with Mg^{++} and with ATP, and it is the formation of the phosphorylated enzyme which leads to an increased requirement for and to a change in the affinity for magnesium.

Results from experiments with g-strophanthin may support the view that there is a shift in the requirement for magnesium when ATP is hydrolyzed.

Magnesium is necessary for the reaction with g-strophanthin[14]. With magnesium, ATP[14,15,16,17,18] and Pi[16,17] increases the reactivity towards g-strophanthin. Sodium has a very pronounced effect

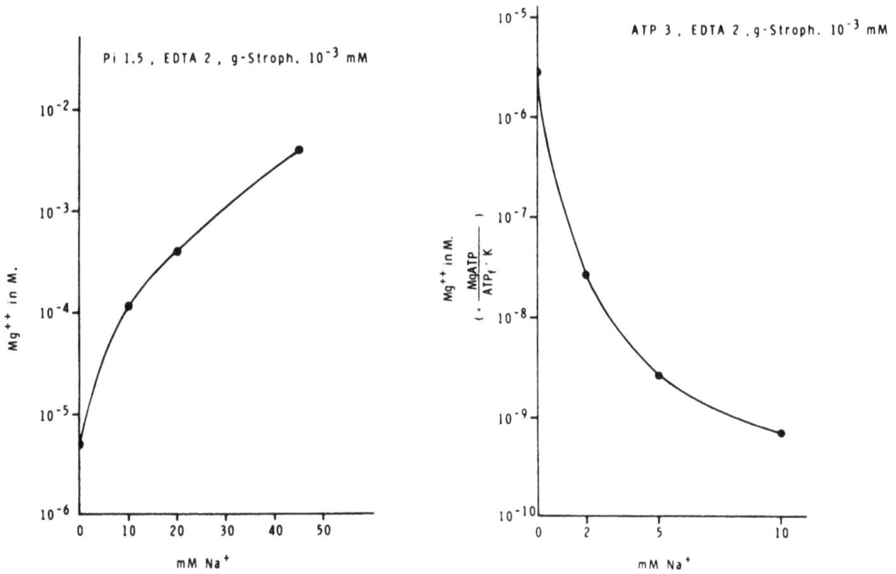

Figs. 1 and 2. The effect of sodium on the concentration of Mg^{++} to give 50% inhibition by 10^{-6} M g-strophanthin, with 1.5 mM Pi, fig. 1, and with 3 mM ATP, fig. 2. Enzyme prepared from ox brain and with a specific activity of 220 umoles Pi/mg protein/h was preincubated for 20 minutes at 37° C in 30 mM tris HCl, pH 7.6, 2 mM EDTA, varying concentrations of magnesium and sodium and with 1.5 mM Pi and 3 mM ATP respectively. After the end of preincubation the hydrolytic activity of the enzyme was tested in a medium with 2 mM EDTA, 5 mM magnesium, 3 mM ATP, 120 mM sodium, 30 mM potassium, and 10^{-6} M g-strophanthin. The initial slope of the curve which gives the hydrolysis of ATP as a function of time was taken to indicate the activity of the enzyme after preincubation. From this the per cent inhibition by g-strophanthin during the preincubation was calculated. The control was enzyme preincubated in the same manner but without g-strophanthin. Correction was made for the activity without sodium plus potassium. For the calculation of Mg^{++} has been used a stability constant of $MgEDTA^{-2} = 10^{-9}\ M^{-1}$, a dissociation constant of $EDTA^{-4} = 10^{-10}$ M[49], and a stability constant of $MgATP^{-2} = 10^4\ M^{-1}$,[50].

on the requirement for magnesium for the reaction with g-strophanthin both with Pi and with ATP, but the effect goes in the opposite direction for the two substrates.

With Pi, sodium increases the requirement for magnesium; this is seen from fig. 1, which shows the concentration of Mg^{++} necessary for 50% inhibition of the enzyme system with 10^{-6} M g-strophanthin and with increasing concentrations of sodium. Potassium has a similar effect as sodium, and the concentration necessary to obtain a certain effect is about 5 times lower than for sodium (not shown). This effect of sodium on the requirement for magnesium for the reaction with g-strophanthin is also found without Pi as substrate.

With ATP, on the other hand, sodium decreases the requirement for magnesium; this is seen from fig. 2, which shows how the concentration of Mg^{++} necessary to give 50% inhibition by 10^{-6} M g-strophanthin decreases when the concentration of sodium is increased. As Mg^{++} equals $MgATP/ATP_f \times k$, the ordinate in fig. 2 also represents the $MgATP/ATP_f \times k$ ratio.

These experiments suggest that there is some kind of competition between sodium and magnesium with Pi as substrate. It is, however, not possible from the present information to tell whether this means a competition between sodium and magnesium for a common site or that sodium at a sodium site has an indirect effect on a magnesium site and vice versa.

With ATP as substrate the competition between sodium and magnesium apparently disappears and the system can react both with sodium and magnesium. It is not possible to tell if it is because both magnesium and sodium shifts the equilibrium towards a g-strophanthin sensitive state that the requirement for magnesium decreases when sodium is increased or it is because sodium has an effect on the affinity for magnesium.

The effect of sodium on the requirement for magnesium with ATP and with Pi is for the inhibition by g-strophanthin, and cannot without further evidence be taken to indicate what happens without g-strophanthin. The rate by which g-strophanthin inhibits the activity of the system is even under optimal conditions low[14,16] while the rate by which the system turns over is high[19]. This may suggest that the effects of Pi, ATP, magnesium, and sodium on the reactivity towards g-strophanthin reflects an effect of these parameters on an equilibrium between a g-strophanthin insensitive and a g-strophanthin sensitive state of the system, and that the effects are also found for the reaction without g-strophanthin in the medium. The qualitative agreement between the effect of sodium on the requirement for magnesium with ATP for the reaction with g-strophanthin, for the phosphorylation[9], and for the ATP-ADP exchange may support this view[12].

The results from the experiments with g-strophanthin agree with the view put forward by Fahn et al.[12] that there is a different requirement for magnesium for different steps in the reaction.

The experiments suggest that the shift in the requirement for magnesium is due to a different way of interaction between sodium and magnesium with and without ATP. With ATP the system reacts both with sodium and magnesium, and the affinity for magnesium is apparently high. The requirement for magnesium for formation of the potassium sensitive phospho-enzyme is higher than for the formation of the ADP-sensitive[13]. This suggests that it is the formation of the ADP-sensitive phospho-enzyme which leads to the shift in the interaction between sodium and magnesium, and by this to the increased requirement for magnesium for the following steps in the reaction.

It suggests the following scheme for the sodium-dependent phosphorylation. Without ATP the system either reacts with sodium or with magnesium; n is a number.

$$nNa^{+} + TS_1 \rightleftharpoons Na_n\text{-}TS_1 \tag{10}$$

$$Mg^{++} + TS_1 \rightleftharpoons Mg\text{-}TS_1 \tag{11}$$

When the system reacts with ATP it is changed in such a way, symbolized by x, that it can react both with sodium and magnesium

$$Na_n\text{-}TS_1 + ATP \rightleftharpoons Na_n\text{-}ATP\text{-}TS_1^{x} \tag{12}$$

$$Na_n\text{-}ATP\text{-}TS_1^{x} + Mg^{++} \rightleftharpoons Na_n\text{-}Mg\text{-}ATP\text{-}TS_1^{x} \tag{13}$$

In the Na_n-Mg-ATP-TS_1^x form the system has catalytic activity and the reaction with ATP leads to the formation of the phosphorylated intermediates. With a low magnesium, the reaction leads to the formation of the ADP-sensitive phospho-enzyme. This phosphorylated intermediate does not react with g-strophanthin[13]. As magnesium seems to be necessary for the reaction with g-strophanthin, this may suggest that this form has lost its magnesium due to the competition between magnesium and sodium which comes at play when ATP is hydrolyzed.

$$Na_n\text{-}Mg\text{-}ATP\text{-}TS_1^{x} \rightleftharpoons Na_n\text{-}P\text{-}TS_1 + Mg^{++} + ADP \tag{14}$$

With a "high" magnesium concentration the reaction leads to the formation of the phospho-enzyme which can react with potassium but not with ADP. It is g-strophanthin sensitive[13], which may suggest that it is a magnesium form of the system. There seems to be no difference in the way the phosphate is bound to the ADP and the sodium sensitive phospho-enzyme[13]. The two phospho-enzymes must therefore differ in some other way. The formation of the oligomycin sensitive phospho-enzyme which seems to be identical with the potassium sensitive phospho-enzyme[20,21] is more sensitive to a decrease in the temperature[21] than the formation of the oligomycin insensitive, ADP sensitive phospho-enzyme[20,21]. This may suggest that the formation of the potassium sensitive phospho-enzyme involves a change in the conformation of the system, TS_1 to TS_2 (cf.[13,21]). The competition between sodium and magnesium, which comes at play when ATP is hydrolyzed, may lead to a decrease in the affinity for both cations:

$$Na_n\text{-}Mg\text{-}ATP\text{-}TS_1{}^x \rightleftharpoons Na_n\text{-}Mg\text{-}P\text{-}TS_2 + ADP \qquad (15)$$

or to a reaction in which magnesium excludes sodium

$$Na_n\text{-}Mg\text{-}ATP\text{-}TS_1{}^x \rightleftharpoons Mg\text{-}P\text{-}TS_2 + nNa^+ + ADP \qquad (16)$$

In the scheme given by Fahn et al.[12] and by Post et al.[13] the ADP and the potassium sensitive phospho-enzymes represent two consecutive steps in the reaction. Another possibility is as shown above that either the one or the other is formed dependent on the magnesium concentration.

The cleavage of the bond between the γ and β phosphate of ATP seems to lead to a change in the internal arrangement of the system and this seems to give a competition between sodium and magnesium, which may mean a change in the way the two ions are bound to the system. The change in the internal arrangement and the change in the interaction between sodium and magnesium may go hand by hand and so to say happen while the bond is cleaved. The change in the way magnesium is bound to the system may be of importance for the rearrangement of the system. With a low concentration, magnesium may be lost from the system during the reaction which may mean that under these conditions the reaction cannot proceed to the formation of the TS_2 state. The result is instead an abortive reaction which does not lead to the formation of TS_2 but to phosphorylation of TS_1 (14).

As mentioned above, it is apparently the same group in the system which is phosphorylated in the ADP sensitive and in the potassium sensitive phospho-enzyme. The phosphate seems to be bound as an acyl phosphate[22,23,24], which means in a bond which normally has a free energy of hydrolysis which is of the same size as for the hydrolysis of the γ-β phosphate bond in ATP.

The phosphorylation of the system is, however, not specific for the reaction with ATP. ITP[11,25], AcP[26,27], pNPP[28] and Pi[29,30] can also phosphorylate the system. The phosphorylation from ITP and AcP is dependent on sodium as the phosphorylation from ATP. The phosphorylation from pNPP and Pi requires a reaction of the system with g-strophanthin and magnesium.

The phosphate from AcP[27], pNPP[28], and Pi[13,30] seems to be bound to the same group on the enzyme as the phosphate from ATP. (It has not been investigated for ITP). The formation of a high energy bond from a relatively low energy substrate as pNPP and from Pi shows that energy for the formation of the bond under these conditions must come from the reaction of the system with g-strophanthin and not from the substrate. The very slow rate of reaction with g-strophanthin suggests that it involves a change in conformation of the system. This may lead to a transformation of "conformational energy" into bond energy and by this to the formation of the high energy bond (see Spiegel et al.[30]).

The formation of the phospho-enzyme with a high energy phosphate bond is thus not specific for a reaction of the system with ATP. There is, however, a specific requirement for ATP for the transport process[31,32]. AcP which has a high energy phosphate bond, which gives a sodium-dependent phosphorylation of the system, and which is hydrolyzed by the system at a rate which is comparable to the rate of hydrolysis of ATP [33] cannot give a transport of sodium[36]. This shows either that formation of the phospho-enzyme is not enough for the transport process or that ATP can phosphorylate under conditions where AcP cannot. In either case it shows that there must be an effect of ATP on the system which precedes the phosphorylation.

The formation of the phosphate bond from low energy sources when the system reacts with g-strophanthin raises the question whether g-strophanthin mimics an effect of sodium plus ATP; or it mimics an effect which is due to the cleavage of the bond between the γ and β phosphate of ATP? In other words, is there an effect of sodium plus ATP as such on the system which leads to the configuration in which the system can react with phosphate and if so, does that mean that the bond between the γ phosphate of ATP and the enzyme system is formed prior to the cleavage of the γ-β phosphate bond? Or is it the cleavage of the bond between the γ-β phosphate which brings the system into a configuration which is comparable to that into which it is brought by g-strophanthin? Or is the conformation which follows from the reaction with sodium plus ATP not identical with that which follows from the reaction with g-strophanthin?

From a comparison of the effect of sodium, potassium and magnesium on the hydrolysis of ATP and of ITP by the system, it was suggested that ATP as such has an effect on the system[35]. This may find support in the observation that ATP plus sodium confers oligomycin sensitivity to the hydrolysis of AcP by the system[36]. This seems to be due to an effect of ATP as such and not to a phosphorylation of the system. The effect is, however, not specific for ATP, it is also obtained by CTP.

The view that ATP as such has an effect on the system may find further support in the observations given above that ATP changes the way the system reacts with sodium and magnesium. It is unknown how specific this effect is. Kinetic studies furthermore suggest that there is a sodium-dependent non-phosphorylated intermediate which precedes the phosphorylation[37].

The changes in the sensitivity towards oligomycin and towards magnesium and sodium by ATP may reflect a change in the system which is a necessary prerequisite for the transformation of the system from $TS_1{}^x$ to TS_2 when the bond between the γ and β phosphate is cleaved. The transformation may follow from this reaction and not from the phosphorylation as such and it may be the transformation which specifically requires an effect of ATP, sodium and magnesium. The phosphorylation may be of importance for stabilizing the system in the TS_2 state until it can react with potassium and by this be dephosphorylated and return to the TS_1 state.

As it seems to be the step which leads to the formation of the potassium sensitive phospho-enzyme which involves a change in configuration, it seems likely that this is the step which specifically requires the reaction with ATP. This would exclude the ADP sensitive phospho-enzyme as an intermediate in the reaction which leads to TS_2, and suggest that it follows from an abortive reaction.

The dephosphorylation by ADP of the "low" magnesium, TS_1, phospho-form suggests that the ΔG for the formation of this phospho-enzyme is close to zero. The lack of reaction of the "high" magnesium form with ADP in spite of a high energy phosphate enzyme bond suggests that formation of the TS_2 phospho-enzyme is an energy-requiring process.

The reaction may be illustrated in a simple way by the following scheme.

"low magnesium"

$$\begin{array}{c} R_2 \\ | \\ TS_1\text{-}R_1 \\ | \\ R_3 \end{array} \xrightleftharpoons{ATP,Na^+,Mg^{++}} \begin{array}{c} R_2 \\ / \\ Na_n\text{-}ATP\text{-}Mg\text{-}TS_1{}^x\text{-}R_1 \\ \backslash \\ R_3 \end{array} \xrightleftharpoons[\searrow Mg^{++}+ADP]{} \begin{array}{c} R_2 \\ | \\ Na_n^+\text{-}TS_1\text{-}R_1 \\ | \\ R_3{\sim}P \end{array}$$

"high" magnesium

$$TS_1\text{-}R_1(R_2)(R_3) \xrightleftharpoons{ATP, Na^+, Mg^{++}} Na_n\text{-}ATP\text{-}Mg\text{-}TS_1{}^x\text{-}R_1(R_2)(R_3) \rightleftharpoons Na_n\text{-}Mg\text{-}TS_2\text{-}R_1\text{-}R_2(R_3{\sim}P) + ADP$$

$$Na_n\text{-}Mg\text{-}TS_2\text{-}R_1\text{-}R_2(R_3{\sim}P) \xrightleftharpoons[mK^+]{} K_m\text{-}TS_1\text{-}R_1(R_2)(R_3) + nNa^+ + Mg^{++} + Pi$$

g-strophanthin

$$TS_1\text{-}R_1(R_2)(R_3) \xrightleftharpoons{Mg^{++},\ G} G\text{-}Mg\text{-}TS_1{}^x\text{-}R_1(R_2)(R_3) \xrightleftharpoons{Pi} G\text{-}Mg\text{-}TS_1{}^x\text{-}R_1(R_2)(R_3{\sim}P)$$

R_1, R_2, R_3 illustrate a conformation inside the system. $R_3 \sim R_2$ is not meant to indicate that there is a high energy bond formed inside the system only to illustrate a change in the distrubution of energy in the system.

According to the view given above it is the reaction which leads to the phosphorylation and not the phosphorylation as such which is specific for the reaction with ATP and which is important to get the system to act as a transport system. It raises a question; is the phosphorylation found with sodium also part of the reaction with sodium plus potassium?

THE EFFECT OF SODIUM PLUS POTASSIUM ON THE PHOSPHORYLATION

The labelling of the system with P^{32} from ATP^{32} is high with magnesium plus sodium, but low with magnesium plus sodium plus potassium (for ref. see[11]). Potassium added to a prephosphorylated system increases the rate of dephosphorylation[9,11,38,39]. This has been taken to indicate that the low labelling with sodium plus potassium is due to an increased rate of dephosphorylation, and that the reaction with sodium plus potassium proceeds via a phosphorylation-dephosphorylation[9,38,39]. The interpretation of the effect of sodium plus potassium on the labelling from the effect of potassium on a prephosphorylated system may, however, depend on the model the system follows.

Models for the Transport Process

Models for the transport process can be divided in two general classes (see Baker and Stone[40]).

1) A model in which the transport of sodium from inside to outside is followed by a transport of potassium from outside to inside, i.e. a system in which the reactions with sodium and potassium follow each other.

2) A model in which the transport of sodium from inside to outside is simultaneous to a transport of potassium from outside to inside, i.e. a system in which the reactions with sodium and potassium is simultaneous.

For convenience the two models have in the following been named a one-unit system for 1 and a two-unit system for 2. These names are not meant to indicate anything about the molecular basis for the models.

In fig. 3 is in a general form given a scheme of a one-unit system (modified from Shaw[41]), and in fig. 4 of a two-unit system. The schemes show the changes in the units which must take place to accomplish an exchange of sodium from inside with potassium from outside of the cell membrane. m and n is the number of cations which is taken up by the units.

Both for the one- and the two-unit system the process can be divided into four steps.

Step 1. One-unit system: a change from the situation in which the system can exchange sodium with the inside solution, i_S, to a situation in which it can exchange sodium with the outside solution, o_S, a translocation (see Mitchell[45]).

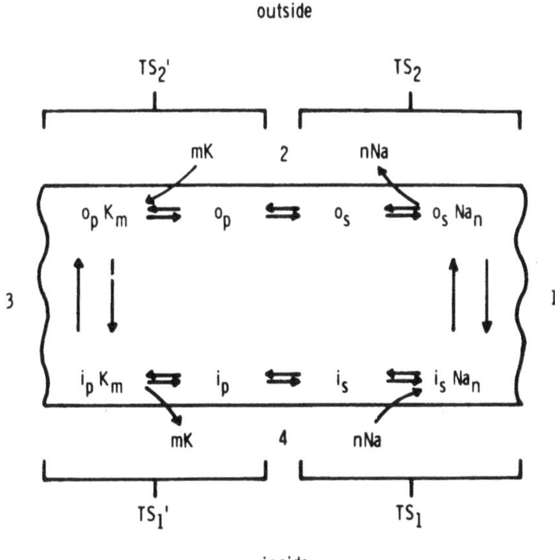

Fig. 3. A general scheme to describe a transport of sodium which is followed by a transport of potassium (modified from Shaw[41]). A one-unit system, see text.

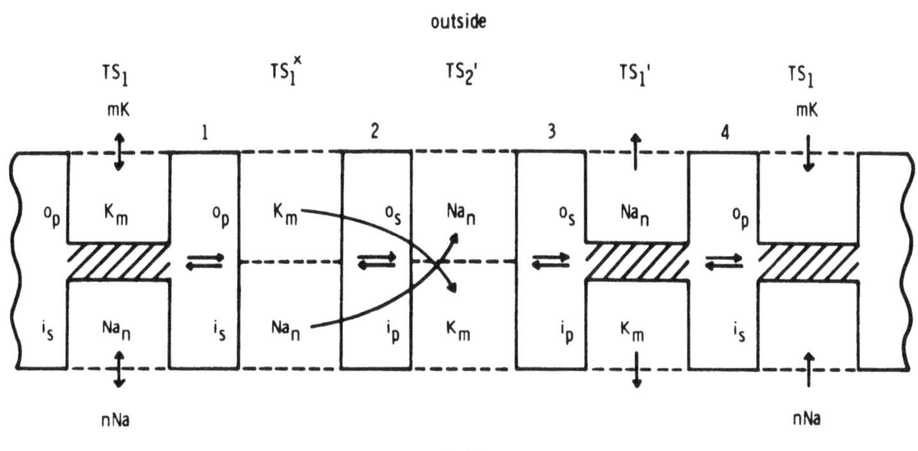

Fig. 4. A general scheme to describe a transport of sodium outwards simultaneously with a transport of potassium inwards, a two-unit system. For explanation, see text.

Two-unit system: a change from the situation in which the o-unit on a potassium form o_p, can exchange potassium with the outside solution, and the i-unit on a sodium form, i_s, can exchange sodium with the inside solution, but not in between each other, to a situation in which the cations under certain conditions (step 2) can be exchanged in between the two units, but to a lower extent with the surroundings. A non-interacted and an interacted state respectively.

Step 2. A change of the affinity of the units.

One-unit system: a change from an affinity for sodium which is higher than for potassium to an affinity for potassium which is higher than for sodium, o_S to o_P (for convenience in the following named a sodium and a potassium affinity, respectively).

Two-unit system: a change of the o-unit from a potassium affinity, o_P, to a sodium affinity, o_S, and for the i-unit from a sodium to a potassium affinity, i_S to i_P.

For both systems it is named a transformation from TS_2 to TS_2'.

Step 3. A back transformation from the TS_2' to the TS_1' state in which the one-unit system can exchange potassium with the inside solution and the two-unit system potassium with the inside and sodium with the outside solution, respectively.

Step 4. A shift in affinity, TS_1' to TS_1. For the one-unit system from a potassium to a sodium affinity, i_P to i_S. For the two-unit system a shift from o_S to o_P and from i_P to i_S.

It is not possible from our present knowledge to decide which of these two classes of models describe the transport process.

<u>One-unit system</u>. In the one-unit system sodium is transported from the inside to the outside, and this is followed by an exchange of sodium for potassium. If it is the same sodium and potassium which triggers the hydrolysis of ATP and which is transported, the effect of potassium on the reaction with ATP must follow that of sodium. This means that the phosphorylation found with sodium in the medium must also be part of the reaction with sodium plus potassium in the medium, and that the effect of potassium is to increase the rate of dephosphorylation. This must be the case whether the phosphate bond is formed prior to or follows from the cleavage of the bond between the γ and β phosphate of ATP.

The concentration of potassium necessary to dephosphorylate the phospho-enzyme is very low relative to the concentration of sodium in the medium, and it increases with the sodium concentration. This shows that the potassium sensitive phospho-enzyme has a site with an affinity for potassium which is much higher than for sodium. This suggests that this form of the system represents the state in which the system can exchange sodium for potassium from outside, i.e. a state in which sodium has been transported from inside to outside.

In the intact cell the transport system can accomplish a Na:Na exchange, and this is inhibited by oligomycin[41]. Oligomycin increases the sodium-dependent labelling from ATP[32], and this seems to be due to a decreased rate of dephosphorylation[11,19,27,43,44]. As oligomycin has no effect or enhances the sodium-dependent ATP-ADP exchange found when the magnesium concentration is low relative to the ATP-concentration[12,21], it seems to be the dephosphorylation of the phospho-enzyme formed with the "high" concentration of magnesium, Na_n-Mg-P-TS_2 which is inhibited by oligomycin and not the dephosphorylation of Na_n-P-TS_1. This suggests that it is the inhibition of the dephosphorylation of Na_n-Mg-P-TS_2 which leads to an inhibition of the Na:Na exchange. This may mean that the steps which precede the formation of Na_n-Mg-P-TS_2 cannot give a Na:Na exchange. It suggests that the system prior to the formation of Na_n-Mg-P-TS_2 is in contact with the inside solution.

It was suggested above that the formation of the ADP sensitive phospho-enzyme was an abortive reaction found when the magnesium concentration was too low, and that this intermediate was not part of the reaction when the magnesium concentration was high enough to give the conformational change of the system from the TS_1 to the TS_2 state, to the potassium sensitive form. If this is correct, it leads to the scheme shown in fig. 5 for the connection between the reaction of the system with ATP and the transport process. In this scheme it is the cleavage of the γ-β phosphate bond of ATP by the system in the form into which it has been brought due to the reaction with ATP, sodium and magnesium that leads to the conformational change and which gives the translocation and the change in affinity from a sodium to a potassium affinity.

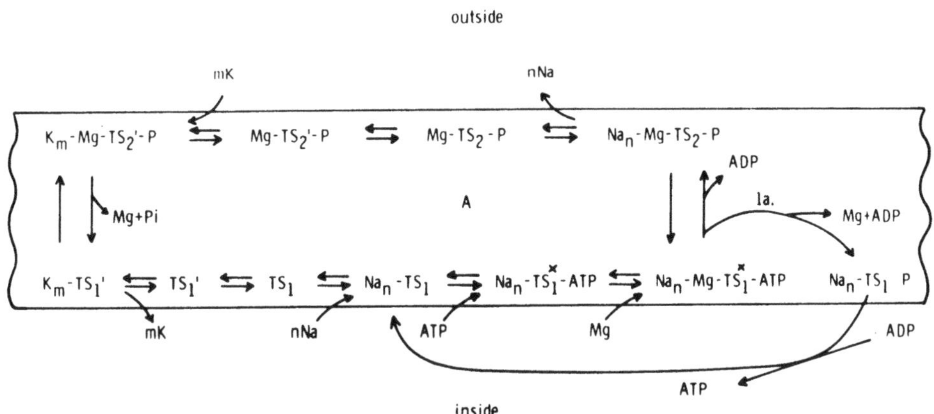

Fig. 5. A one-unit model for the transport process. For explanation, see text.

As discussed above, the formation of a phospho-enzyme bond with a free energy of hydrolysis which is of the same size as for the hydrolysis of the phosphate bond in ATP does not necessarily mean that the ΔG for the reaction which leads to the formation of the phospho-enzyme, TS_2, is zero. It seems most likely that this step is energy-requiring.

In the scheme shown in fig. 5 the chemical change is intimately related to the translocation and the change in the affinity from a sodium to a potassium affinity (see Mitchell[45]). For a detailed discussion of the scheme see Skou[14].

It must be emphasized that a translocation from TS_1 to TS_2 in a one-unit system not necessarily means a macroscopie (relative to the dimensions of the membrane) movement of a carrier molecule from the inside to the outside of the membrane and that phosphate is moved from inside to outside. The system may both in the TS_1 and TS_2 state be in contact with the inside of the membrane, but in the TS_1 state be able to exchange cations with the internal solution, and in the TS_2 state with the external solution, as for example in an alternating gate system[46].

Two-unit system. In the scheme shown in fig. 6 each of the two units in the non-interacted state exists in an equilibrium between a potassium form and a sodium form. For the o-unit the apparent affinity for potassium is 60-100 times the apparent affinity for sodium[9,10], which means that the equilibrium for the o-unit is towards the potassium form, o_p. For the i-unit the apparent affinity for sodium is 4-6 times the apparent affinity for potassium[10], which means that the equilibrium for this unit is towards the sodium form, i_s.

Assuming that no hybrid forms exist (see however Middleton[47]) there are four combinations of the transport system, a Na_o/Na_i, a K_o/K_i, a K_o/Na_i, and a Na_o/K_i form (see fig. 6). Calculations on the basis of the apparent affinities and on basis of the sodium and the potassium concentrations in the external and internal solutions of mammalian cells give about 30-40% of the system on the K_o/Na_i form, 30-40% on the K_o/K_i form, 10-15% on the Na_o/Na_i form, and 10-15% on the Na_o/K_i form in an intact cell membrane. The ratio between the forms varies with the sodium/potassium ratio in the internal and external solutions respectively.

With sodium and no potassium in the medium, there is sodium at both units, i.e. the system is on a Na_o/Na_i form (see fig. 6). It is unknown whether the effect of sodium, which leads to the phosphorylation, is due to sodium being on the i-unit or on both units, but it seems likely that it is sodium on the i-unit which is important. Potassium increases the rate of dephosphorylation of a prephosphorylated system in concentrations which are so low com-

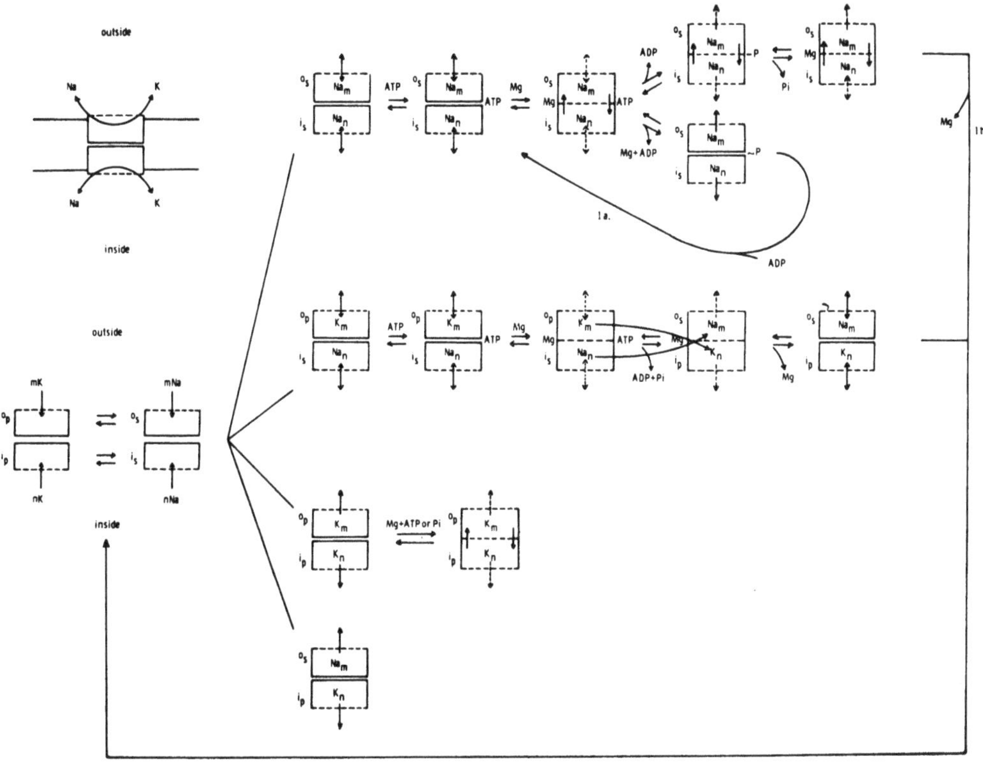

Fig. 6. A two-unit model for the transport process. For explanation, see text.

pared to the concentration of sodium in the medium that the effect must be on a site with an affinity for potassium which is higher than for sodium, i.e. on the o-unit[9,11]. With the system on the K_o/Na_i form, there is then at the same time an effect of Na_i which gives a high rate of cleavage of the bond between the γ and β phosphate of ATP and of K_o which gives a high rate of hydrolysis of the enzyme-phosphate bond.

If the phospho-enzyme is formed between enzyme and ATP prior to the cleavage of the bond between the γ and β phosphate and this leads to transformation of the system into the catalytic active state, formation or no formation of a phosphorylated enzyme with the system on the K_o/Na_i form depends on the ratio between the rate constants for the cleavage of the γ-β bond and the enzyme-γ bond. This means that the phosphorylation found with sodium but no potassium needs not necessarily be an intermediate in the reaction with sodium plus potassium.

This is neither the case if the formation of the bond between the enzyme and the phosphate with the system on the Na_o/Na_i form follows from the cleavage of the bond between the γ and β phosphate of ATP. The increased rate of dephosphorylation when potassium is added to a prephosphorylated enzyme may be an indication that the enzyme-phosphate bond cannot exist, i.e. cannot be formed when potassium is in the medium, i.e. with the system on the K_o/Na_i form.

In the experiments with ATP^{32} it is found that even if the labelling with sodium plus potassium in the medium is decreased compared to the labelling without potassium, there is left a certain but low labelling which decreases when ATP is completely hydrolyzed[11]. This suggests that the phospho-enzyme is formed also with sodium plus potassium. It is, however, difficult to exclude that part of the system with sodium plus potassium in the medium under the experimental conditions used has been on the Na_o/Na_i form (cf. fig. 6), and that the labelling in the medium is due to this.

In this model it is assumed that the reaction of the system on the Na_o/Na_i and K_o/Na_i forms with ATP and magnesium leads to a change in the interaction between the two units; it is shown as a change from a non-interacted to an interacted state, TS_1 to $TS_1{}^x$ in fig. 4. In the interacted state the system has the catalytic activity and on the K_o/Na_i form the cleavage of the γ-β phosphate bond of ATP leads to a change in conformation of the system in such a way that the o-unit is changed from a potassium to a sodium affinity, o_p to o_s, and the i-unit from a sodium to a potassium affinity, i_p to i_s, with a following exchange of the cations in between the units. The reaction may or may not proceed via formation of a phospho-enzyme. In fig. 6 it is shown without formation of a phospho-enzyme.

If no phosphorylated intermediate is formed, the TS_2 state is a transition state which only exists so to say while ATP is hydrolyzed, and in which the exchange of the cations is intimately connected to the hydrolysis of ATP.

On the Na_o/Na_i form the reaction leads to the formation of either the one or the other of the two phospho-enzymes dependent on the magnesium concentration (see fig. 6).

The different requirement for the formation of the two phospho-enzymes formed with sodium in the medium[13] cannot in the two-unit model explain why there is a different requirement for magnesium for the sodium-dependent ATP-ADP exchange and the sodium plus potassium-dependent ATP hydrolysis. The phospho-enzymes formed with sodium are on a pathway which is different from the pathway for the hydrolysis with sodium plus potassium. A way of explaining the different requirement for magnesium for the two pathways is that the system on the Na_o/Na_i form has a higher affinity for Mg^{++} than on the K_o/Na_i form; this may be due to a different way of interaction of the two units on the Na_o/Na_i and the K_o/Na_i form when the system reacts with ATP.

A different pathway for the hydrolysis of ATP on the Na_o/Na_i and on the K_o/Na_i form with different affinities for Mg^{++} and ATP explains why the same system seems to behave as two different systems, one found with sodium and no potassium, the other found with sodium and potassium[48]. (For a detailed discussion of the two-unit model see Skou[14]).

CONCLUSION

It seems not possible from our present knowledge to draw any final conclusions about the role of the phosphorylated intermediates in the transport process.

ABBREVIATIONS

ATP, adenosine triphosphate; ADP, adenosine diphosphate; ITP, inosine triphosphate; CTP, cytosine triphosphate; AcP, acetyl phosphate; pNPP, p-nitrophenyl phosphate; Pi, inorganic phosphate; ATP_f, uncomplexed ATP; Mg^{++}, uncomplexed magnesium.

REFERENCES

1. Skou, J. C., Physiol. Rev. 45:596 (1965).
2. Heinz, E., A. Rev. Physiol. 29:21 (1967).
3. Albers, R. W., Ann. Rev. Biochem. 36:727 (1967).
4. Glynn, I. M., Br. Med. Bull. 24:165 (1968).
5. Glynn, I. M., J. Physiol. (Lond.) 160:18P. (1962).
6. Laris, P. C. and Letchworth, P. E., J. Cell. Comp. Physiol. 60:229 (1962).
7. Whittam, R., Biochem. J. 84:110 (1962).
8. Garrahan, P. J. and Glynn, I. M., J. Physiol. (Lond.) 192:217 (1967 d).
9. Post, R. L., Merritt, C. R., Kinsolving, C. R., and Albright, C. D., J. Biol. Chem. 235:1796 (1960).
10. Skou, J. C., in "The Molecular Basis of Membrane Function" (D. Tosteson, ed.), pp. 455-482, Prentice Hall, Inc., New Jersey (1969).
11. Skou, J. C. and Hilberg, C., Biochim. Biophys. Acta 185: 198 (1969).
12. Fahn, S., Koval, G. J., and Albers, R. W., J. Biol. Chem. 241:1882 (1966).
13. Post, R. L., Kume, S., Tobin, T., Orcutt, B., and Sen, A. K., J. Gen. Physiol. 54:306 s (1969).

14. Skou, J. C., Butler, K., and Hansen, O., Biochim. Biophys. Acta .....
15. Matsui, H. and Schwartz, A., Biochim. Biophys. Acta 151: 655 (1968).
16. Albers, R. W., Koval, G. F., and Siegel, G. F., Mol. Pharmacol. 4:324 (1968).
17. Sen, A. K., Tobin, T., and Post, R. L., J. Biol. Chem. 244:6596 (1969).
18. Tobin, T. and Sen, A. K., Biochim. Biophys. Acta 198: 120 (1970).
19. Bader, H., Post, R. L., and Bond. E., Biochim. Biophys. Acta 150:41 (1968).
20. Fahn, S., Koval, G. J., and Albers, R. W., J. Biol. Chem. 243:1993 (1968).
21. Blostein, R., J. Biol. Chem. 245:270 (1970).
22. Hokin, L. E., Sastry, P. S., Galsworthy, P. R., and Yoda, A., Proc. Nat. Acad. Sci. U.S.A. 54:177 (1965).
23. Nagano, K., Kanazawa, T., Mizuno, N., Tashina, Y., Nakao, T., and Nakao, M., Biochem. Biophys. Res. Comm. 19:759 (1965).
24. Bader, H., Sen, A. K., and Post, R. L., Biochim. Biophys. Acta 118:106 (1966).
25. Schoner, W., Beusch, R., and Kramer, R., European J. Biochem. 7:102 (1968).
26. Bond, G. H., Bader, H., and Post, R. L., Fed. Proc. 25: 567 (1966).
27. Israel, Y. and Titus, E., Biochim. Biophys. Acta 139:450 (1967).
28. Interrusi, C. E. and Titus, E., Mol. Pharmacol. 6:99 (1970).
29. Lindenmayer, G. E., Langhfer, A. H., and Schwartz, A., Arch. Biochem. Biophys. 127:187 (1968).
30. Siegel, G. F., Koval, G. F., and Albers, R. W., J. Biol. Chem. 244:3264 (1969).
31. Hoffman, J. F., Fed. Proc. 19:127 (1960).
32. Hoffman, J. F., Circulation 26:1201 (1962).
33. Bader, H. and Sen, A. K., Biochim. Biophys. Acta 118:116 (1966).
34. Mullins, L. F. and Brinley, F. J., J. Gen. Physiol. 53: 704 (1969).
35. Skou, J. C., Biochim. Biophys. Acta 42:6 (1960).
36. Askari, A. and Koyal, D., Biochem. Biophys. Res. Comm. 32:227 (1968).
37. Saito, M., in "Ann. Report Biol. Works", pp. 15-47, published by Faculty of Science, Osaka Univ. (1969).
38. Ahmed, K. and Judah, J. D., Biochim. Biophys. Acta 104:112 (1965).
39. Charnock, J. S. and Potter, H. A., Archiv. Biochem. Biophys. 134:42 (1969).
40. Baker, P. F. and Stone, A. J., Biochim. Biophys. Acta 126:321 (1966).
41. Shaw, T. J., Ph.D. Thesis, Cambridge University (1954).

42. Garrahan, P. J. and Glynn, I. M., J. Physiol. (Lond.).
43. Whittam, R., Wheeler, K. P., and Blake, A., Nature 203: 720 (1964).
44. Gruener, N. and Avi-Dor, Y., Biochem. J. 100:762 (1966).
45. Mitchell, P., Advances in Enzymology. (Nord, F. F., ed.), vol. 29, pp. 33-87, Intersci. Publ., New York (1967).
46. Jardetzky, O., Nature (London) 211:969 (1966).
47. Middleton, H. W., Arch. Biochem. Biophys. 136:280 (1970).
48. Neufeld, A. H. and Levy, H. M., J. Biol. Chem. 244:6493 (1969).
49. Sillén, L. G. and Martell, A. E., Stability Constants of Metal Ion Complexes. Special publication No. 17, London, The Chemical Society (1964).
50. Nørby, J. G., Acta Chem. Scand. (1970). In press.

FUNCTIONS OF PHOSPHOLIPIDS IN ADENOSINTRIPHOSPHATASES ASSOCIATED WITH MEMBRANES

A.Bruni, A.R. Contessa and P. Palatini

Institute of Pharmacology, University of Padova

35100 Padova (Italy)

Summary

Similarly to other enzyme complexes structurally associated with membranes, sodium, potassium-stimulated ATPase and mitochondrial ATPase require phospholipids for full activity. Two functional aspects reflecting the need for phospholipids in these organized systems will be discussed. The first is dealing with the influence of individual phospholipids on the components of the enzyme complexes. The second is concerning the interaction with external lipid-soluble molecules, inhibitors included. Evidences are summarized illustrating the antagonism between phospholipids and oligomycin, inhibitor of both mitochondrial and sodium, potassium-stimulated ATPases.

The removal of oligomycin effect by external phospholipids may provide useful indications on the mechanisms involved in the interaction between this antibiotic and the enzyme complexes.

The purpose of this paper is to discuss some results concerning the function of phospholipids in the activity of enzyme complexes associated with cellular membranes. Among these, notable examples are the ATPase (1) required in the terminal sequence of the mitochondrial oxidative phosphorylation and the ATPase (2) catalyzing the active extrusion of sodium and uptake of potassium through the plasma membrane of the cell. The first is commonly referred as mitochondrial ATPase and the second is known as sodium and potassium-stimulated ATPase or, simply, transport ATPase.

The role of phospholipids in the activity of membrane-located enzymes has been the subject of many studies in the recent years and excellent reviews are available for general informations (3, 4,5).

Components of Particulate Mitochondrial and Transport ATPases

Particulate mitochondrial and transport ATPases have in common several properties (a) both are firmly bound to a membranous structure. (b) They can couple the hydrolysis of ATP to the transport of cations across the membrane in which they are located. (c) Under favourable conditions they can generate ATP from ADP and inorganic phosphate. This was shown recently by Garrahan and Glynn (6) for transport ATPase. (d) They need phospholipids for full complement of activities. (e) When included in the membrane both are sensitive to the inhibitor oligomycin (7,8).

In spite of these analogies, there are not evidences showing similarities in the components of the two ATPases. The main reason is that detailed studies on transport ATPase have been hampered by the difficulty to get this enzyme complex in a soluble and purified form.

Uesugi et al. and Medzihradsky et al. (9,10), using the nonionic detergent Lubrol, succeded in the isolation from calf and guinea-pig brains of an apparently soluble fraction with ouabain sensitive, sodium and potassium-stimulated ATPase activity; this fraction had a molecular weight of 670,000.

However, polyacrylamide gel electrophoresis of the preparation (11), after removal of lipids and solubilization with fenol, urea and acetic acid, revealed 12 proteic bands.

The heterogeneity of preparations with sodium and potassium-stimulated ATPase activity is illustrated by the experiment reported in Fig. 1. A partially purified preparation of transport ATPase from calf heart was solubilized with sodium dodecyl sulfate plus urea and subjected to polyacrylamide gel electrophoresis. Among several minor bands, two major components were seen. One migrated exactly as phosphorylase a, the other was a diffuse band of lower molecular weight. An apparently purer preparation, "solubilized" by deoxycholate from rabbit renal cortex was recently reported (12).

On the contrary, particulate mitochondrial ATPase is the best characterized among the membrane-associated ATPases. Systematic work in Racker's laboratory (1,13) led to isolation and purification of the enzyme responsible for the splitting of ATP. The

FIG. 1

POLYACRYLAMIDE GEL ELECTROPHORESIS OF A PREPARATION FROM CALF HEART CONTAINING THE Na^{+} - K^{+} - STIMULATED ATPase.

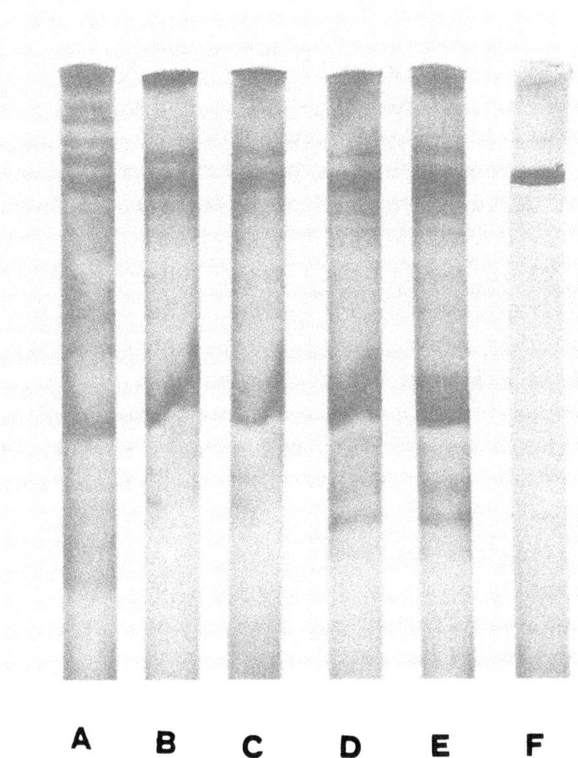

A B C D E F

The method of Matsui and Schwartz (33), with some modification (34), was used. Phospholipids were extracted with aqueous aceton and the residue subjected to electrophoresis according to Weber and Osborn (35).

A. Enzyme complex at the stage of dexycholate extraction.

B. The same after treatment with NaI.

C and D. Fractions obtained after treatment of NaI purified enzyme with sonic oscillations.

E. F + D.

F. Phosphorylase a.

component, named coupling Factor 1, is a soluble protein of molecular weight of 280,000.

Subsequent attempts to reconstitute the membrane-associated oligomycin sensitive ATPase resulted in the identification of several components needed for the full complement of ATPase activity (14-17).

Requirement for Phospholipids

The essential role of phospholipids in the activation of particulate mitochondrial ATPase and in the restoration of oligomycin sensitivity is well documented (14-17). Activation of the enzyme complex is produced either by a crude preparation of phospholipids of different origin or by several individual phospholipids.

No similar clear results were presented on transport ATPase; although several reports (18-21) indicate the requirement for phospholipids by this enzyme complex, the nature of the phospholipids involved is a question far from resolved.

Some evidences (19-21) indicate the essential role of phosphatidylserine in the activation of this ATPase. However, the addition of phosphatidylserine to phospholipid-depleted preparations made by extraction with deoxycholate (22) resulted in a low specific activity, giving the impression that in the full active enzyme some other phospholipid might be involved. Recently (12), one of these preparations was further purified but was stimulated only two-fold by phosphatidylserine.

In view of the cation binding property by negatively charged phospholipids and the apparently specific activating effect on transport ATPase by phosphatidylserine, it has been proposed (21) that this phospholipid takes directly active part in the movement of cations across the cell membrane.

Role of Phospholipids in the Organization of Enzyme Complexes

Data on the function of phospholipids in the organization of enzyme complexes such as the membrane associated ATPases are mainly derived from studies on the resolution and reconstitution of mitochondrial inner membrane (23). In this context it may be profictable to make a comparison between the effect of phospholipids on two mitochondrial enzyme systems which were successfully resolved and reconstituted: the oligomycin-sensitive ATPase (14-17) and the TTB*-sensitive succinate-ubiquinone reductase (24).

*TTB = 4,4,4-trifluoro-1-(2-thienyl)-1-3-butanedione.

The reconstitution process gives the possibility to analyze the differences in the activity among individual phospholipids (ref. 17 and 24). Cardiolipin, known to be effective in preserving the activity of succinate dehydrogenase (25), activated the mitochondrial ATPase but elicited low oligomycin sensitivity and was completely inactive for the reconstitution of ubiquinone-reductase activity. Bovine phosphatidylethanolamine was fully active in the stimulation of reconstituted succinate-ubiquinone reductase and of oligomycin-sensitive ATPase. Soya bean lecithin stimulated much better mitochondrial ATPase and restored high oligomycin sensitivity. Bovine lecithin greatly stimulated the succinate-ubiquinone reductase activity, but had a limited affect on the mitochondrial ATPase and on oligomycin sensitivity.

The differences among purified phospholipids are of interest considering that both succinate-ubiquinone reductase and oligomycin sensitive ATPase belong to the same cellular membrane.

In the evaluation of the effectiveness of phospholipids in restoring the oligomycin sensitivity, some difficulty arises from the observation that these components have antagonistic effect on the oligomycin-induced inhibition (26).

Antagonism between Phospholipids and Oligomycin

The partecipation of phospholipids in the interaction between lipid-soluble molecules and membrane-linked enzymes has been suggested (3,24). This function of phospholipids in a membrane may be relevant also for the effect of inhibitory compounds. Cerletti et al. (27) have reported that pretreatment with phospholipase C decreased the inhibitory effect of TTB* on particulate preparations with succinate-dichlorophenol-indophenol reductase activity.

Moreover, the possibility of removing bound oligomycin from submitochondrial particles, washing with a phospholipid-containing medium, is also known (15).

Recently (26), it was observed that the simple addition of a crude preparation of phospholipids to the incubation medium prevented the oligomycin-induced inhibition of both transport and mitochondrial ATPases. The protective effect could be reproduced using individual purified phospholipids. This observation prompted an investigation on the structural requirement for the phospholipids effect.

Table I shows data on the antagonism by purified lecithins.

---

*TTB = 4,4,4-trifluoro-1-(2-thienyl)1-3-butanedione.

TABLE I

ANTAGONISM BETWEEN OLIGOMYCIN AND LECITHINS

Additions	Oligomycin-induced inhibition	
	Mitochondrial ATPase	Transport ATPase
None	100	100
0.3 mg bovine brain lecithin	42	51
0.5 mg bovine brain lecithin	24	40
0.3 mg egg lecithin	52	51.5
0.5 mg egg lecithin	42.7	39
0.3 mg soya bean lecithin	62	-
0.5 mg soya bean lecithin	-	57
0.5 mg rat-liver mitoch.lecithin	5	54.5
0.5 mg synth. dipalmitoyl lecithin	40.5	-

Transport ATPase was a preparation from calf heart (see Fig.1,D) with specific activity of 15-25 μmoles ATP split/mg protein/h at 37°C. Ouabain inhibition was 95-97 %. Mitochondrial ATPase was the activity of submitochondrial particles from rat-liver (Kielley and Bronk (36)),80-100 μmoles ATP split/mg protein/h at 37°C. Incubation medium: 100 mM NaCl, 20 mM KCl, 2.5 mM $MgCl_2$, 50 mM TRIS.HCl,pH 7.4, 3.0 mM ATP.TRIS,pH 7.4, 25 mM sucrose, 0.1 mM EDTA; final volume, 1.0 ml; 20 min at 37°C.
Submitochondrial particles, 30 μg proteins (oligomycin concentration 1.6 μg/mg protein); transport ATPase, 130-250 μg proteins (oligomycin concentration 24-46 μg/mg protein).
The inhibition by oligomycin in the absence of phospholipids (80-90% with submitochondrial particles and 60-70% with transport ATPase) was taken as 100; the values in the presence of phospholipids represent the residual effect of oligomycin.
Phospholipids were "solubilized" by sonic oscillations.

Bovine brain lecithin and, expecially, mitochondrial lecithin were effective antagonists of oligomycin but they were more active on particulate mitochondrial ATPase than on transport ATPase. Egg lecithin produced the same antagonism with both ATPases. Synthetic dipalmitoyl lecithin, tested at 37° which is very close to its temperature transition, was rather effective, whereas soya bean lecithin exhibited less activity. These results suggest the following conclusions: (a)phospholipids with uncharged polar portion can be effective antagonist of oligomycin; (b) there are differences among the various lecithins indicating the influence of the hydrocarbon chains; (c) the different behavior of mitochondrial lecithin in the two enzymatic preparations suggests that the affinity between the added phospholipids and those present in the enzyme complexes can be important. The possibility to be exchanged with the phospholipids of the fragmented mitochondria, as shown by Fleischer and Brierley (28), is presumably an explanation of the high activity of mitochondrial lecithin in the case of mitochondrial ATPase. This indicate that active oligomycin is bound to phospholipids in the particles and can be removed through the exchange.

The same possibility is supposedly essential when external phospholipids have to remove the oligomycin already bound to the membrane fragments by a pretreatment procedure. Bovine brain lecithin can reactivate a preparation of transport ATPase inhibited by a pretreatment with the antibiotic (26).

In Table II it is seen that the antagonism by bovine brain lecithin is much more effective than the limited reversal induced by the addition of serum albumin . This latter compound did not influence the removal of oligomycin effect produced by the lecithin.

In Table III it is shown that the antagonism between oligomycin and bovine brain lecithin is influenced by the temperature of incubation: it was more evident at 37° in comparison to 26°.Although the ionic composition did not produce relevant effects using submitochondrial particles, the antagonism was better seen upon addition of sodium and potassium.

Table IV shows the effect of other phospholipids. With the remarkable exception of bovine brain sphingomyelin on mitochondrial ATPase, they were at best slightly active. Phosphatidylserine, whose partecipation in the operation of transport ATPase was recently proposed, is included in this group. Both phosphatidylserine and sphingomyelin produced inhibitory effect on the ATPase preparations: the former inhibited more the mitochondrial ATPase, the latter more the transport ATPase. The mechanism of

TABLE II

DIFFERENTIAL EFFECT OF SERUM ALBUMINE AND BOVINE BRAIN LECITHIN ON THE OLIGOMYCIN-INDUCED INHIBITION OF MITOCHONDRIAL ATPase

Additions	µmoles ATP split/mg protein/h at 37°		
	Without oligom.	With oligom.	% inhib.
None	92	11.4	87.6
0.5 mg lecithin	88.5	69.5	21.5
1.0 mg bovine serum album.	88.5	19	78.5
lecithin + albumine	85	73.5	13.5

Experimental conditions as described in Table I. ATPase activity of submitochondrial particles from rat-liver. Oligomycin concentration was 1.6 µg/mg protein.

TABLE III

EFFECT OF TEMPERATURE OF INCUBATION ON THE ANTAGONISM BETWEEN BOVINE BRAIN LECITHIN AND OLIGOMYCIN

Temp.	Lecithin (0.5 mg)	µmoles ATP split/mg protein/h at 37°		
		Without oligom.	With oligom.	% inhib.
26°C	-	77	7.7	90
"	+	77	42	45.5
37°C	-	88.5	11.5	87
"	+	96	65	32

Experimental conditions deseribed in Table I. ATPase activity of submitochondrial particles from rat-liver. Oligomycin concentration was 1.6 µg/mg protein.

TABLE IV

ANTAGONISM BETWEEN OLIGOMYCIN AND VARIOUS PHOSPHOLIPIDS

Additions	Oligomycin-induced inhibition	
	Mitochondrial ATPase	Transport ATPase
None	100	100
0.5 mg bov.brain phosphatidyl-ethanolamine	80	82
0.5 mg soya bean phosphatidyl-ethanolamine	80	83.5
0.2 mg bov.brain phosphatidyl-serine	-	86
0.2 mg bov.brain sphingomyelin	52.5	-

Experimental conditions described in Table I.

these inhibitory effects deserves further investigations.

The effectiveness of sphingomyelin in preventing the oligomycin-induced inhibition of mitochondrial ATPase confirms the activity of phospholipids which form in water uncharged aggregates.

The role of hydrophobic fatty acid chains was studied using simpler lipids. In Table V it is seen the antagonism by triglycerides and monoglycerides. Among the former, triacetin was inactive but the antagonism was very evident as soon as the fatty acid chains became longer. Tributyrin was very effective and some activity was still present in trioctanoin. Among triglycerides containing saturated fatty acid chains, the effectiveness decreased as the chain length increased; on the contrary maximum activity was reached using triglycerides with unsaturated members (triolein); essentially the same results were seen with monoglycerides among which only monolein produced some antagonism.

Frome these results it can be concluded that the antagonism

TABLE V

ANTAGONISM BETWEEN OLIGOMYCIN AND LIPIDS

Additions	Oligomycin-induced inhibition of mitochondrial ATPase
None	100
(a) Triglycerides	
0.5 mg Triacetin	106
0.5 mg Tributyrin	24
0.5 mg Trioctanoin	80
0.5 mg Trilaurin	100
0.5 mg Tripalmitin	102
0.5 mg Tristearin	93
0.5 mg Triolein	0
(b) Monoglycerides	
0.5 mg Monoacetin	104
0.5 mg Monopalmitin	98
0.5 mg Monoolein	88
0.5 mg Monostearin	101

Experimental conditions as described in Table I; ATPase activity of submitochondrial particles from rat-liver. Oligomycin concentration was 1.6 μg/mg protein. Lipids (10 mg/ml) were suspended in 0.25 M sucrose, 10 mM TRIS.HCl, pH 7.4, 1.0 mM EDTA and briefly (2-3 min) sonicated, keeping the temperature close to 20°C.

with oligomycin is shown by neutral lipids provided the paraffin residues are not short enough to produce water soluble compounds and not long enough to induce a very high increase in the melting point or in the paraffin residues are present unsaturated fatty acids. The best configuration seems to be represented by triolein.

All the active lipids are liquid at room temperature and form in water, after sonication, a dispersed liquid phase in which oligomycin can presumably be dissolved. This situation is similar to the antagonism between antimycin and ubiquinone observed by Takemori and King (29) and suggests similar properties between the two antibiotics. The binding to the phospholipid portion of the membrane as mean to reach the site concerned in the catalytic activity could explain some of the results, discussed by Kaniuga et al. (30) on the inhibitory effect produced by antimycin on mitochondrial respiratory chain.

Concluding Remarks

Two functional aspects reflecting the need for phospholipids in membrane fragments with enzymatic activity were discussed. The first is concerning the organization of these multicomponents assemblies, the second the interaction with external molecules.

Besides the well known (31,32) property to promote the formation of a membrane, which is a rather aspecific effect shared by many of these compounds, phospholipids have a direct influence on the single constituents of a membranous enzyme complex.

Example of inhibition by phospholipids is the effect of cardiolipin on the reconstitutive property of purified cytochrome b (24); example of stabilizing effect is that of acidic phospholipids on succinate dehydrogenase (25); example of activation is that of phosphatidylethanolamine on succinate-ubiquinone reductase (24) and oligomycin-sensitive mitochondrial ATPase (17).

The different behavior of individual phospholipids offers an explanation to the need for several phospholipids in a membrane organization and suggests that these compounds can not be randomly distributed in a membrane.

The property of phospholipids to provide a suitable medium for the interaction with lipid-soluble molecules can be reflected by the antagonism with oligomycin (26), a phenomenon which shows specificity. The following requirements for such an antagonism were established: (a) the lack in the phospholipid ag gregates of a net charge at physiological pH; (b) the affinity with the phospholipids of the enzyme complex; (c) the presence of unsaturated fatty acids in the paraffin chains.

The evidences obtained so far allow to consider two mechanisms for an explanation of the antagonism between oligomycin and phospholipids. The formation of appropriate aggregates upon addition of phospholipids to the incubation medium, constitutes a system which effectively competes for oligomycin with the phospholipids of the enzyme preparation. The competition can be so effective to induce a redistribution of oligomycin even when the antibiotic is already bound to the enzyme complex. In addition, the exchange between the added and the phospholipids present in the membrane fragments, when it is possible, may increase the effectiveness of the antagonism. Both possibilities suggest that oligomycin finds in the phospholipid portion of the membrane the suitable environment to be dessolved and in this way to reach te context of the membrane, with subsequent modification of the functional part. It is likely that a reversible equilibrium can be established between the amount which remains bound to the phospholipids and that able to join the site involved in the catalytic activity.

These considerations are also relevant from a pharmacological point of view. Many drugs are lipid-soluble compounds, which influence membrane associated activities. The possibility of a specific interaction between drugs and phospholipids indicates that the lipid part of a membrane not only becomes a site for drug effect but can be considered as a mean to channel the drug to its final target. In this way pharmacological effects can be regulated by the composition of the lipid part of the membrane.

References

1.Pullman,M.E.,Penefsky,H.S.,Datta,A.,and Racker,E., J.Biol.Chem. 235:3322 (1960).

2.Skou,J.C., Biochim.Biophys.Acta, 23:394 (1957).

3.Green,D.E.,and Fleischer,S.,in "Biochemical problems of lipids" (A.C.Frazer,ed.) p.325.Elsevier Pub.,Amsterdam (1963).

4.vanDeenen,L.L.M., in "Regulatory Functions of Biological Membranes" (J.Järnefelt,ed.) p.72. Elsevier Pub., Amsterdam (1968).

5.Tria,E.,and Barnabei,O., in "Structural and Functional Aspects of Lipoproteins in Living Systems" (E.Tria and A.M.Scanu,eds.) p. 144. Academic Press, New York (1969).

6.Garrahan,P.J.,and Glynn,I.M., J.Physiol. (London) 192:237 (1967).

7.Lardy,H.A.,Johnson,D.,and McMurray,W.C., Arch.Biochem.Biophys., 78:587 (1958).

8.Glynn,I.M., Biochem.J., 84:75P (1962).

9.Uesugi,S.,Kahlenberg,A.,Medzihradsky,F.,and Hokin,L.E., Arch.

Biochem.Biophys., 130:156 (1969).
10.Medzihradsky,F.,Kline,M.K.,and Hokin,L.E., Arch.Biochem.Biophys. 121:311 (1967).
11.Hokin,L.E.,J.General Physiol., 53;327s (1969).
12.Towle,D.W.,and Copenhaver,J.H.,Jr., Biochim.biophys.Acta, 203:124 (1970).
13.Penefsky,H.S.,Pullman,M.E.,Datta,A.,and Racker,E.,J.Biol.Chem. 235:3330 (1960).
14.Kagawa,Y.,and Racker,E., J.Biol.Chem., 241:2461 (1966).
15.Kagawa,Y.,and Racker,E., J.Biol.Chem., 241:2467 (1966).
16.Bulos,B.B.,and Racker,E., J.Biol.Chem.,243:3891 (1968).
17.Bulos,B.B.,and Racker,E., J.Biol.Chem.,243:3901 (1968).
18.Ohnishi,T.,and Kawamura,H., J.Biochem. (Tokyo) 56;377 (1964).
19.Fenster,L.J., and Copenhaver,J.H.,Jr.,Biochim.Biophys.Acta, 137:406 (1967).
20.Tanaka,R.,J.Neurochem.,16:1301 (1969).
21.Wheeler,K.P.,Whittam,R.,Nature, 225:449 (1970).
22.Tanaka,R.,and Strickland,K.P.,Arch.Biochem.Biophys., 111:583 (1965).
23.Racker,E.,and A.,Bruni, in "Membrane Models and the Formation of Biological Membranes" (L.Bolis and B.A.Pethica,eds.) p.138. North-Holland Pub., Amsterdam (1968).
24.Bruni,A.,and Racker,E., J.Biol.Chem., 243:962 (1968).
25.Cerletti,P.,Giovenco,M.A.,Giordano,M.G.,Giovenco,S.,and Strom R., Biochim.Biophys.Acta, 146:380 (1967).
26.Palatini,P.,and Bruni,A., Biochem.Biophys.Res.Comm. (in press).
27.Cerletti,P.,Giovenco,M.A.,Giordano,M.G.,Caiafa,P., and Magni,G., in "Abstract of Communications presented at 4th Meeting of FEBS" p. 145 (1967).
28.Fleischer,S.,and Brierley,G., Biochim.Biophys.Acta, 53:609 (1961).
29.Takemori,S.,and King,T.E., J.Biol.Chem., 239:3546 (1964).
30.Kaniuga,Z.,Bryla,J.,and Slater,E.C., FEBS Symposium,Volume 19, p.285 (1969).
31.McConnell,D.G.,Tzagoloff,A.,Mac-Lennan,D.H., and Green,D.E., J.Biol.Chem., 241:2373 (1966).
32.Kagawa,Y.,and Racker,E., J.Biol.Chem., 241:2475 (1966).
33.Matsui,H.,and Schwartz,A., Biochim.Biophys.Acta,128:380 (1966).
34.Contessa,A.R.and Bruni,A., (manuscript in preparation).
35.Weber,K., and Osborn,M., J.Biol.Chem., 244:4406 (1969).
36.Kielley,W.W.,and Bronk,J.R., J.Biol.Chem., 230:521 (1958).

LIPOPROTEIN INTEGRITY AND ENZYMATIC ACTIVITY OF THE ERYTHROCYTE MEMBRANE

B. Roelofsen, R.F.A. Zwaal and L.L.M. van Deenen

Department of Biochemistry

University of Utrecht, The Netherlands

In the last few years, a variety of methods for solubilizing proteins from erythrocyte stroma have appeared in the literature. Nearly all depend on: i) detergent action, in some cases combined with acidic or alkaline pH; ii) disruption by organic solvent systems; or iii) a combination of both, e.g. lipid extraction by organic solvents, followed by solubilization of the residual protein by detergents.

The harshness of methods involving detergents, compounded by their ability to bind both lipids and proteins, may result in a significant alteration of the protein from its native state. On the other hand treatment with organic solvent systems may also lead to partial or complete denaturation of the protein fraction.

In view of this, it was deemed important to compare the recovery of specific activities of different enzymes, particularly the membrane bound ATPase, after treatment of human erythrocyte ghosts with organic solvents and detergents.

Solubilization of the Erythrocyte Membrane by n-Butanol and n-Pentanol.

A method for the extraction of lipids from enzymes of animal tissues by n-butanol was developed by Morton (1) in 1955. This method was adapted to the ghost of ox erythrocytes by Maddy (2,3). He succeeded in solubilizing 90-95% of the membrane proteins in water. The butanol fractionation procedure of Maddy has been applied to human and other mammalian erythrocyte ghosts (Zwaal and van Deenen (4), Zwaal (5), Rega et al. (6,7)) giving also high recoveries (80-90%) of the membrane proteins in the butanol-saturated water phase.

The butanol fractionation method has the advantage over other techniques, of extracting lipid and in the same step solubilizing the erythrocyte membrane proteins in water. Since butanol is also used in the extraction of lipids from enzymes, it can reasonably be assumed that extensive denaturation of the proteins does not occur.

The butanol treatment was applied to human erythrocyte ghosts, which were prepared by a minor modification of the method of Dodge et al. (8), as described by Zwaal (4). The haemoglobin-free ghost suspensions obtained, were frozen overnight at -15°C, thawed the next day and washed three times with boiled ditilled water at 2°C. The water washed stroma suspensions were mixed with equal volumes of butanol at 0°C for 25 min. After centrifugation at 30,000 x g the mixtures were separated by a thin, insoluble interfacial film into an upper butanol layer containing the majority of the lipids (90-95%) and a water layer mainly containing the membrane apoproteins (85-90%) (Fig. 1).

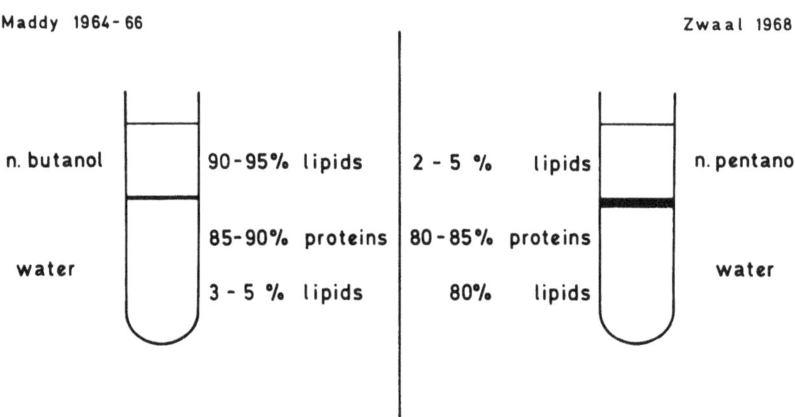

Fig. 1 Percent recovery of proteins and lipids in organic phase and water phase after the treatment of aqueous ghost suspensions with n-butanol and n-pentanol respectively (9).

It should be noted that in order to obtain the yields mentioned, freezing of the ghost suspensions overnight, followed by three successive washings with ion-free water seemed to be essential steps in the isolation procedure. As long as the system was not ion-free, a large amount of insoluble protein was found at the butanol-water interface.

Besides n-butanol, the treatment of aqueous ghost suspensions was carried out with n-pentanol under identical conditions. It appeared that n-pentanol also brought about the solubilization of erythrocyte ghosts, but in contrast to the butanol procedure, both the proteins and the lipids were recovered in the aqueous phase, yields being: 80-85% of the proteins and about 80% of the lipids (including cholesterol) in the water phase, and only 2-5% of the phospholipids in the pentanol phase (Fig. 1).

With mixtures of n-pentanol and n-butanol, the solubilization of ghosts led to a recovery of phospholipid in the water phase that was dependent on the ratio of pentanol and butanol (5,9) (Fig. 2).

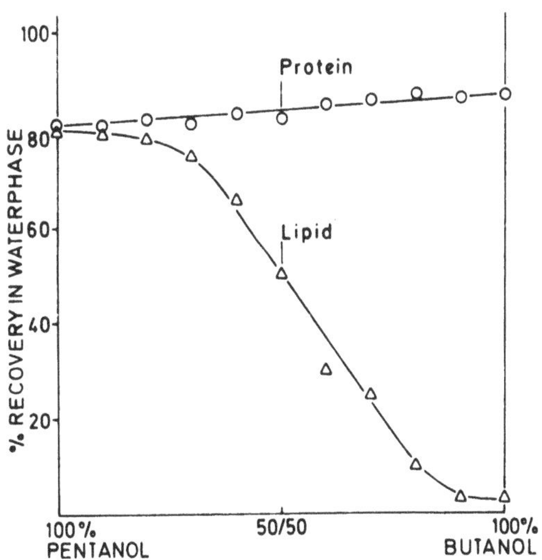

Fig. 2 Protein and phospholipid recovery in the water phase after treatment with pentanol-butanol mixtures of aqueous ghost suspensions (9).
0-0, protein;Δ-Δ , phospholipid.

Whereas the water phase after butanol treatment contains the apoproteins from the red cell membrane, although still containing 3.6% phospholipids and 0.9% cholesterol by dry weight, the solubilized material in the water phase after pentanol treatment consists of lipoprotein structures rather than a mixture of discrete lipids and proteins.

Support for the existence of lipoprotein structures in the aqueous phase after pentanol treatment has been obtained from proton magnetic resonance studies (10); electron microscopy, which shows a homogenous structure; and by centrifugation in sucrose gradients (5,9). Centrifugation in a density gradient at pH 5.5 indicated the presence of two major fractions, both consisting of lipid and protein, and no distinct lipid layer on top of the gradient was detectable (Fig. 3). The lipid

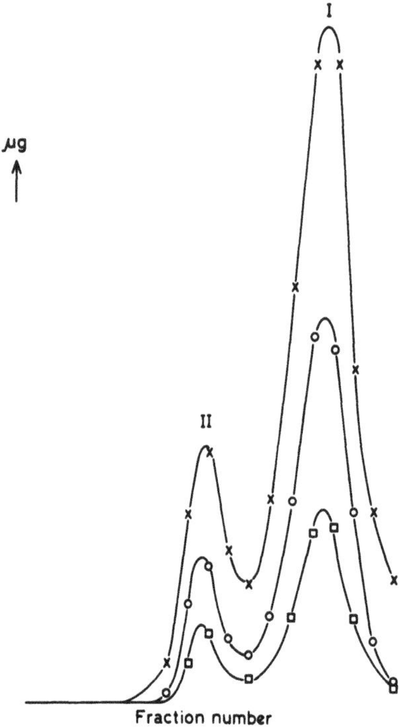

Fig. 3 Fractionation of lipoproteins solubilized from human erythrocyte ghosts by n-pentanol treatment (9). Centrifugation in a sucrose gradient (13-55%). X-X, protein; O-O, phospholipid; □ - □, cholesterol.

composition of the pentanol layer was identical to that of the original ghost, and no differences could be detected in the phospholipid-protein ratio, or in the phospholipid composition of the two fractions obtained by the density gradient centrifugation. In a control experiment, the protein-containing water phase obtained after butanol treatment was mixed with an aqueous lipid dispersion prepared by sonication of total erythrocyte lipids, prior to the centrifugation procedure. In this case, after centrifugation the lipid remained floating at the top, while the apoproteins had moved into the gradient.

Apoproteins and lipoproteins prepared by butanol and pentanol treatment respectively, showed some similarities. Both of them remained soluble after heating the solutions for 1 h at 100°C, but after freezing and thawing or lyophilization, the proteins were converted into an insoluble form, probably due to irreversible denaturation.

The recovery of specific activities of different enzymes in the water layers after pentanol and butanol extraction of aqueous ghost suspensions was compared. Enzymes were selected whose activities could be detected in reasonable amounts in the ghost suspension, prior to the extraction procedure with butanol and pentanol. It has previously been shown that not all the enzymes present in the erythrocyte are bound to the membrane with the same tenacity (11).

In table I, the percentage recoveries of specific enzymatic activity are given for 7 enzymes after pentanol and butanol extraction.

The recovery of aldolase activity, measured according to Rutter et al. (12) was found to be somewhat lower in the apoproteins than in the lipoproteins. On the other hand, the glyceraldehyde-3-P-dehydrogenase activity, measured according to Allison (13), was found to be higher in the apoproteins with regard to the lipoproteins and the original ghost. These results suggest that both pentanol and butanol cause minimal changes in the conformation of these enzymes. Similar conclusions can be drawn with respect to 3-P-glycerate kinase, especially when it is taken into account that 25% of the activity is lost during the dialysis procedure after butanol and pentanol extraction. Thus the lower recovery of 3-P-glycerate kinase compared to the other two glycolitic enzymes cannot only be attributed to the butanol or pentanol treatment of ghost.

No acyl-CoA:monoacylphosphoglyceride acyltransferase activity, measured as described by van den Berg (14), could be

TABLE I

PERCENT RECOVERY OF SPECIFIC ENZYMATIC ACTIVITY AFTER BUTANOL OR PENTANOL TREATMENT OF AQUEOUS GHOST SUSPENSIONS FROM HUMAN ERYTHROCYTES

Ghost = 100%

	Apoproteins	Lipoproteins
Aldolase	69- 74	89-112
Glyceraldehyde-3-P-dehydrogenase	103-110	76- 96
3-P-glycerate kinase	57- 64	54
Acyl-CoA: monoacylphosphoglyceride acyltransferase	0	0
ATPase (Mg^{++})	0	0
ATPase (Na^+, K^+)	0	0
Acetylcholinesterase	1- 3	20- 27

detected in the apoproteins or in the lipoproteins. It is not known whether the complete loss of activity is due to denaturation of the acyltransferase or to the removal of lipids by the organic solvents.

The acetylcholinesterase activity, measured by a minor modification of the procedure developed by Michel (15), was found to be practically lost in the apoproteins whereas about a quarter of the original activity could be detected in the lipoproteins. It is not known whether or not this enzyme requires a lipoprotein structure for its activity, although it has been proposed that acetylcholinesterase serves as a good indicator of membrane stability (16,17). Bearing this in mind, it may be speculated that acetylcholinesterase activity in the lipoproteins after pentanol extraction indicates that at least part of the integrity of the lipoprotein complexes in the original membrane has been preserved.

The Mg^{++}-dependent and the (Na^{+} + K^{+})-stimulated ATPase activities (measured by a minor modification of the procedure developed by Post et al. (18)) also did not withstand the butanol and pentanol extraction. Similar observations have been reported by Rega et al. (7) after the butanol fractionation of human red cell ghosts. It has been argued that the ATPase system requires the integrity of lipoprotein complexes in the erythrocyte membrane (19). The loss of ATPase activity in the apoproteins may be due primarily to the extraction of lipids by butanol, whereas during the pentanol treatment not only a minor extraction of lipids but also a possible change in the lipid-protein organization may be responsible.

ATPase Activities of Ghosts after Treatment with Subsaturated Concentrations of Butanol and Pentanol, and with Detergents.

The influence of low concentrations of butanol and pentanol on the ATPase activities of human red cell membranes was investigated. These investigations were carried out on erytrocyte ghosts prepared by the method of Dodge et al. (8)*. In addition, ghosts were prepared by a minor modification (20) of the procedure of Parpart (21), using CO_2-saturated water to haemolyse the erythrocytes as well as for washing of the ghosts**.

* to be denoted as HF-ghost (HF = haemoglobin-free).
** to be denoted as CO_2-ghost.

The ghost suspensions were mixed with 0-16% (v/v) butanol, or with 0-12% (v/v) pentanol* and allowed to stand for 20 min at 4°C prior to the ATPase estimation procedures.

Fig. 4 represents the percentage specific activities of the Mg^{++}-dependent and (Na^+ + K^+)-stimulated ATPase as a function of the butanol and pentanol concentration. As far as the HF-ghost is concerned, it appeared that upon increasing the amount of butanol and pentanol, the (Na^+ + K^+)-ATPase first becomes significantly activated (Fig. 4, upper part), and then shows a sharp decrease in activity. The Mg^{++}-dependent ATPase activity does not exhibit an initial increase, but starts to fall off right away (Fig. 4, lower part), indicating that this enzyme is much more sensitive to the action of the present alcohols. With respect to the ghost suspension prepared by the method of Parpart, the treatment with butanol caused only a small, and possibly not significant, initial activation of both of the ATPase activities.

In addition to butanol and pentanol, the influence of some detergents has been studied. Increasing amounts of Na-DOC, Triton X-100 and SDS were added to aqueous ghost suspensions and the mixtures were allowed to stand for 20 min at room temperature, prior to the ATPase incubation. As Fig. 4 indicates, the three detergents tested caused, at least qualitatively, the same effects as butanol and pentanol, i.e. at low concentrations a significant increase of the (Na^+ + K^+)-ATPase of the HF-ghost, and only a small possibly not significant activation of both of the ATPases of the CO_2-ghost.

It has to be noted that there is no agreement in literature with respect to the effect of detergents on the (Na^+ + K^+)-activated ATPase. It has been reported that low concentrations of detergents activate this enzyme (Somogyi (23)), Marchesi and Palade (24), Jørgensen and Skou (25)), which is in agreement with our own observations using HF-ghosts. On the other hand, Israel (26) reported an inhibition of the (Na^+ + K^+)-ATPase, even at very low detergent concentrations. From our experiences, it may be suggested that those differences can be due to differences in the procedures by which cell membranes have been prepared.

* Solubility in water at 4°C: Butanol 12% (v/v)
Pentanol 4,7% (v/v) (22)

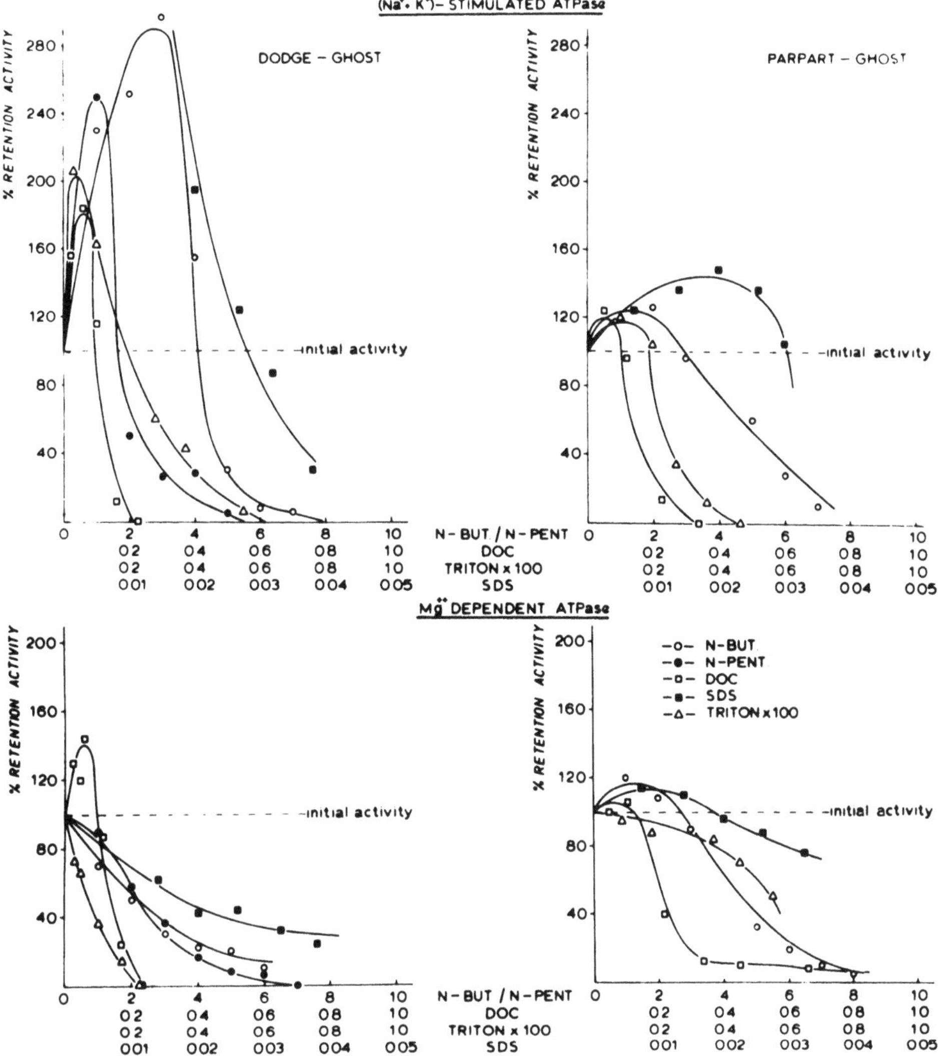

Fig. 4 Percent retention of specific activity of (Na^+ + K^+)-stimulated, and Mg^+-dependent ATPase of human erythrocyte ghosts prepared according to Dodge et al. (8) (left panels), as well as prepared by the method of Parpart (21) (right panels). Concentrations of the added compounds are expressed in percentages, butanol and pentanol (v/v), detergents (w/v).

The observed activation of the transport ATPase of the HF-ghost by low concentrations of butanol, pentanol and detergents, may be explained by some small disruption of the membrane structure by which the accessibility of the enzyme for its substrate is increased. The highest specific activity of the ($Na^+ + K^+$)-ATPase of the HF-ghost, obtained by the treatment with the present alcohols or detergents, reaches the same level as the specific activity of the untreated CO_2-ghost, which one is two to three times as high as the specific activity of the untreated ghost prepared from the same erythrocyte pool by the method of Dodge et al. This difference in behaviour by the two ghost preparations may be explained by the fact that the osmotic shock, undergone by erythrocytes, is much less severe in Dodge's procedure as compared to Parpart's. One could imagine that the drastic osmotic shock in Parpart's method causes a disruption of the membrane structure which is, at least with respect to the ATPase system, comparable with the structural changes of the HF-ghost, as they are induced by low concentrations of butanol, pentanol or detergents.

When after reaching the point of maximal activation of the ($Na^+ + K^+$)-ATPase, the butanol-, pentanol-, or detergent concentration is further increased, the ATPase activities are decreased, which is accompanied by a solubilization of the ghosts. Upon complete solubilization of the membranes, neither of the two ATPase activities could be detected any more, which may indicate more drastic changes in the lipoprotein structure or the lipid-protein binding.

Other organic solvents besides n-butanol and n-pentanol have been tested for their influence on the ATPase activities present in human erythrocyte ghosts.

ATPase Activities of Ghosts after Lyophilization and Partial or Complete Removal of Lipids.

A nonuniformity in the binding of lipids can be demonstrated in lyophilized cell membranes by means of solvent extractions (19,20,27). Ether treatment of the dehydrated ghosts extracts all of the membrane sterol and a defined part of the phospholipids denoted as "loosely bound lipid". Subsequent extraction with more polar solvents removes the remaining phospholipids considered to be "strongly bound" (Fig. 5). All clases of phospholipids are present in both lipid fractions, but quantitatively, the ethanolamine-containing ones appeared to dominate in the loosely bound fraction. This differentiation is not brought about by a difference in solubility, and the fatty acid composition of the phospholipids present in both fractions is very similar (20). Lyophilization of ghosts,

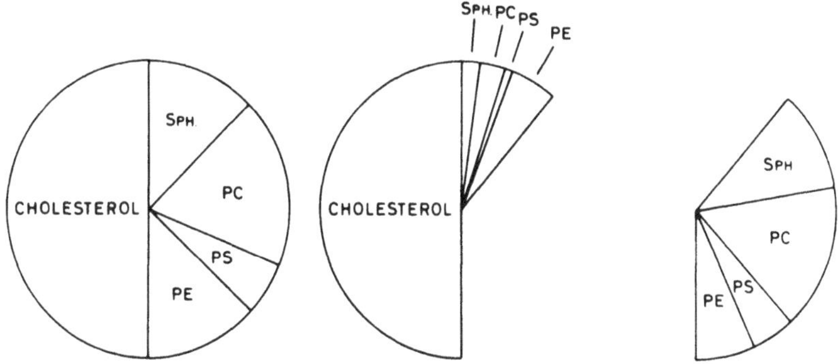

Fig. 5 Scheme for the extraction of "loosely" and "strongly" bound lipids from human erythrocyte ghosts and the relative distribution over the corresponding fractions (19).
Abbreviations: SPH = sphingomyelin; PC = phosphatidylcholine; PS = phosphatidylserine; PE = ethanolamine containing phosphoglycerides.

prepared by the method of Parpart, and successive removal of the loosely bound lipid fraction appeared not to alter the ATPase activities. Subsequent extraction of the strongly bound phospholipids, or enzymatic cleavage by means of phospholipases, caused a complete loss of both of the ATPase activities (Fig. 6). These results suggest that the phospholipids, not removed by ether treatment of freeze-dried ghosts, are involved in these processes. However, it is very doubtful whether the complete loss of the ATPase activities is primarily due to the removal of lipids by the extraction of red cell ghosts with ethanol-ether (3:1, v/v) or acetone (90 vol %). Although extraction with 90 vol % of acetone has been succesfully used for the delipidation of enzyme systems (Lester and Fleischer (28)), a denaturation of the ATPase protein cannot be excluded.

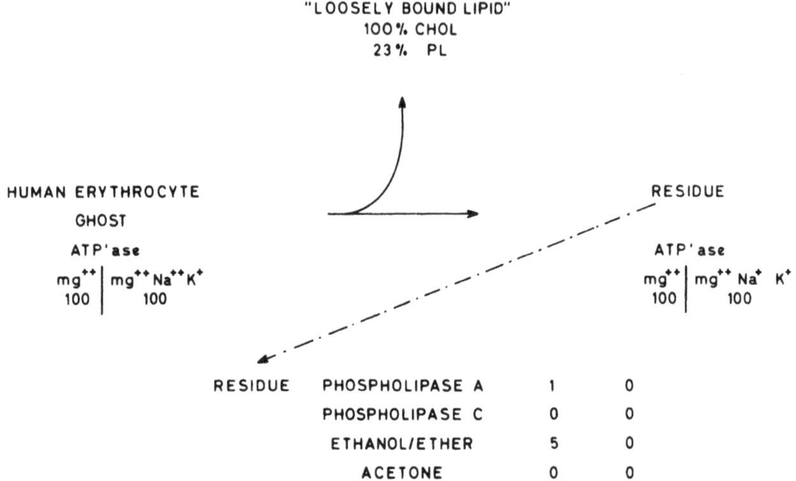

Fig. 6 Percent retention of ATPase activities after anhydrous ether extraction of human erythrocyte ghosts, and after successive treatment with pancreatic phospholipase A or phospholipase C (Bac. cereus), as well as after extraction with ethanol-ether (3:1, v/v) or acetone (90%) (30).

Effects of Organic Solvent - Water Mixtures on the ATPase Activities of Human Erythrocyte Ghosts.

Our previous investigations (19,29) have demonstrated that a number of organic solvents mixed with water and under isotonic conditions, cause lysis of the erythrocyte, at a sharply defined concentration for each solvent. The haemolytic action did not correlate with the concentration of the solvent but with the dielectric constant of the medium. It was suggested that at a critical value of the dielectric constant, essential bonds between lipids and proteins are broken (29).

Pretreatment of CO_2-ghosts with organic solvents, present at concentrations 3 per cent lower than the haemolytic concentration, did not significantly influence the ATPase activities (19,30) (Fig. 7). On the other hand, treatment of the ghosts with aqueous mixtures containing organic solvents 2 per cent above the haemolytic concentration gave practically complete inactivation (Fig. 7).

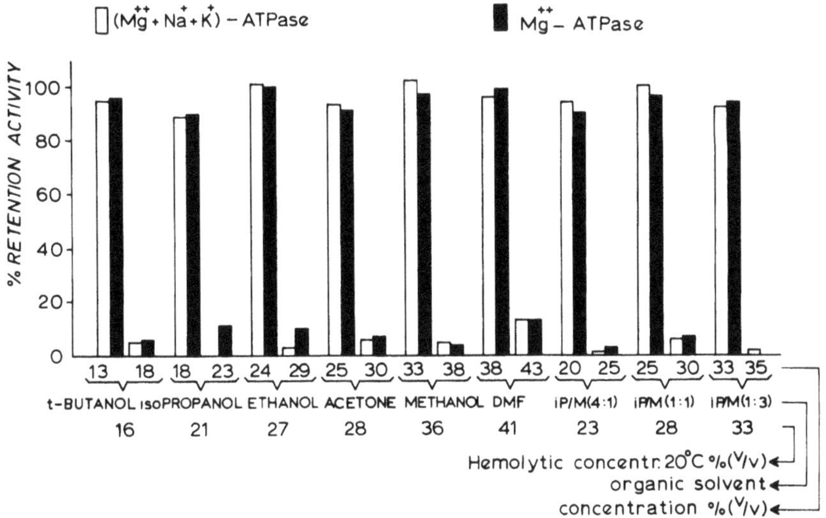

Fig. 7 Percent retention of ATPase activities after pretreatment of human erythrocyte ghosts with organic solvent - water mixtures (30). Open and solid bars represent (Mg^{++} + Na^{+} + K^{+})-ATPase and Mg^{++}-ATPase activities, respectively.
Abbreviations: DMF = dimethylformamide; iP = iso-propanol; M = methanol.

Mixtures of methanol and iso-propanol in water induced inactivation at total concentrations of organic solvent, which agreed with the theoretical values derived by making the assumption that the dielectric constant alters linearly with the composition of the system (29).

It has to be emphasized that none of the organic solvents present at concentrations mentioned in Fig. 7 extracted any detectable amount of lipid material from the ghosts. The loss of ATPase activities, induced by concentrations of organic solvents similar to those causing haemolysis, may support the suggestion made before, that at these critical concentrations of the organic solvents essential bonds between the (strongly bound) phospholipids and the proteins are weakened, altered, oreven broken. However, it must be emphasized that prior to the ATPase incubation, the organic solvents had been removed by washing of the ghosts. Removal of the solvents by lyophilization or by a more gradual procedure such as dialysis against an aqueous buffer, did not restore the ATPase activities either. It may be speculated that after these treatments of the ghosts, the phospholipids do not occupy their initial binding sites on the membrane proteins.

This might support the suggestion already made by Schatzmann (31), that the major role of the phospholipids involved in the ATPase activity, could be the maintenance of the correct structure of the enzyme, which is essential for its action. In that case, one could suggest that the structure of the membrane proteins and in particular that of the ATPase enzyme system, is irreversibly changed by the treatment of the ghost with aqueous mixtures of organic solvents present at their critical concentrations.

Although the acyl-CoA: monoacylphosphoglyceride acyltransferase, whose activity was found to be completely lost in the lipoproteins prepared by the pentanol treatment, did fully withstand the last mentioned organic solvent treatments, one still has to consider the previous conclusion with some reserve, as long as it has not been established that the ghost treatments under discussion do not primarily attack the protein moiety.

Restoration of ATPase Activity by Adding Back Phospholipids, after Pretreatment of Ghosts with Phospholipases or Detergents.

The involvement of phospholipids, or the lipoprotein nature of the membrane bound ATPase enzyme system, has been reported by a variety of investigators during the last ten years. At first this involvement has been demonstrated by a reduction of the ATPase activity after treatment of the enzyme preparation with phospholipase A (Tatibana (30)) or phospholipase C (Schatzmann (31)).

However, as the phospholipases used in these studies have been rather impure preparations (snake venom or bacterial culture filtrate), almost certainly being not free of proteolytic enzymes, the reported loss of activity did not deliver strong evidence for the involvement of phospholipids. Therefore, several investigators working on this field have tried to restore the ATPase activity by adding back phospholipids (33-40). Although there was no or little agreement in literature on the nature of the lipid bringing about a reactivation, it turned out that in this respect phosphatidylserine has to be considered as the most active one. However, as has been mentioned already by Hegyvary and Post (39), in most cases the specific activity was low, and the reactivation small. These authors postulated that in order to obtain better evidence for the involvement of phospholipids in the (Na^+ + K^+)-ATPase, one should demonstrate: "1) a loss of phospholipids from the membrane preparation as a result of inactivation treatment and restoration of the same lipids following reactivation, and 2) a constant ratio of the phospholipid content to the specific activity of the membrane preparation". Those starting points may be debatable, for they imply that all lipid molecules present in a membrane perform the same function. As has been mentioned before, it could be demonstrated that the removal of the "loosely bound" phospholipids (including about 25% of the phosphatidylserine molecules) from the ghosts by anhydrous ether extraction, did not alter the ATPase activities at all.

However, in the same article Hegyvary and Post demonstrated an inactivation of the (Na^+ + K^+)-ATPase from kidney membrane, which was proportional to the digestion of the phospholipids by phospholipase A (Naja Naja snake venom). Reactivation could be obtained by the addition of a commercial soybean extract, asolectin. It was emphasized by Hegyvary and Post that the ATPase activity only could be restored as long as the enzyme has been partially inactivated.

As far as the reactivation by addition of phospholipids reported by most of the other investigators is concerned, it is obvious that also in those cases a reactivation has been obtained on basis of an enzyme that had been partially inactivated. Of course, some discretion has to be observed, for some authors have presented their data in such a way that it is impossible to find out whether any remaining activity had been present or not. The reason why we like to emphasize this point will become clear from the following.

In contrast to other investigators, who have been working with rather impure phospholipases, we have used for the treatment of erythrocyte ghosts highly purified phospholipase A (a globular protein of M.W. 13,800, isolated from porcine pancreas), which causes hydrolysis of the fatty acid ester linkage at the C_2 position of the phosphoglycerides (41). This treatment resulted in a complete breakdown of the ethanolamine, choline and serine containing phosphoglycerides, while the Mg^{++}-dependent as well as the (Na^+ + K^+)-stimulated ATPase activities were found to be completely lost. We tried to restore the activity by adding back ultrasonically dispersed solutions of lipids, such as total lipids obtained from fresh human erythrocytes and chromatographically pure phosphatidylserine, isolated from pig brain by the method described by Sanders (42). Under all conditions that we tried we could not obtain any reactivation. Our inability to reactivate the ATPase is in agreement with the observations of Hegyvary and Post (39). The most straight forward evidence for the involvement of phospholipids in the ATPase enzyme system would be a restoration of the ATPase activity upon addition of phospholipid to an enzyme preparation which had been completely inactivated by the treatment with pure phospholipase A. Unfortunately such experimental results have not been obtained.

Israel (26) demonstrated that the (Na^+ + K^+)-stimulated ATPase from untreated beef brain microsomes could be activated by the addition of phospholipids to the ATPase incubation medium. It was found that of the phospholipids used, only phosphatidylserine (PS) activated the enzyme. Also Wheeler and Whittam (40) reported that the (Na^+ + K^+)-ATPase activity present in a membrane containing fraction of ox brain, was raised by phosphatidylserine in the ordinary enzyme preparation, before it had been treated with other agents. This effect is explained by the authors by the suggestion that some phospholipid may be lost during the preparation of the enzyme. This suggestion is not confirmed by our observation that removal of the "loosely bound" phospholipids does not alter the ATPase activities present in the erythrocyte membrane.

The activating effect of phosphatidylserine on the (Na^+ + K^+)-stimulated ATPase could be confirmed by our own experiments, using human erythrocyte ghosts. The Mg^{++}-dependent ATPase activity of ghosts did not exhibit any significant alteration by the addition of increasing amounts of PS, regardless whether the ghosts have been prepared according to the procedure of Dodge et al. or the method of Parpart. The (Na^+ + K^+)-stimulated ATPase, present in ghosts prepared according to Dodge et al., showed an activation with increasing amounts of PS, reaching a maximum of about 3 times as high as the original activity, at a ratio of mg ghost protein to mg PS being about 1:1. The maximum activity of the transport ATPase brought about by the addition of the optimal amount of PS to the HF-ghost, reached the same level as the corresponding activity present in ghosts prepared according to Parpart, whose transport ATPase activity was slightly or not altered upon addition of PS. This finding seems to be comparable with the effect which detergents exhibit on the (Na^+ + K^+)-stimulated ATPase present in both types of ghost preparation, although the detergents inactivated the Mg^{++}-ATPase in both preparations, which is in contrast to the observations made upon the addition of phosphatidylserine.

As will be clear from the foregoing, one has to be very careful with the interpretation of experiments concerning the reactivation of the membrane bound ATPase upon addition of PS or other lipids. As long as there is some remaining activity, one has to take into consideration that the observed increase of activity upon the addition of phospholipids, might be caused rather by an activation of the remaining activity than a reactivation of the activity which has been lost. In this respect it has to be emphasized that several reactivation experiments published, have been carried out by using a partially inactivated ATPase preparation.

Nevertheless, Tanaka (36,38,43) and co-workers seemed to be successful in the demonstration of a reliable reactivation of the (Na^+ + K^+)-stimulated ATPase by the addition of phospholipids. The enzyme was prepared from cerebral cortex of beef cattle by treatment with deoxycholate. After some further purification, a soluble preparation has been obtained, showing no (Na^+ + K^+)-ATPase activity without phospholipids. Reactivation of the enzyme could be demonstrated upon addition of acidic phospholipids, such as phosphatidic acid, phosphatidylinositol and phosphatidylserine, whereas neutral lipids such as lecithin and phosphatidylethanolamine were found to be inactive. In order to investigate the molecular structure of phospholipid essential to activate the (Na^+ + K^+)-stimulated ATPase, Tanaka and Sakamoto (43) have tested a series of compounds. Besides of the acidic phospholipids, mono- and diacyl phosphates were found to be effective in this respect. It is suggested by the authors that the essential structures needed for activation are a phosphate group plus one or two fatty acyl residues. From these findings of Tanaka and Sakamoto, one has to conclude that there is not an absolute requirement of PS for the (Na^+ + K^+)-ATPase to accomplish its function.

Although it might be true that within the cell membrane PS plays a specific role in the activity of the ATPase enzyme system, it can be replaced by other compounds having a negatively charged group in combination with one or two hydrophobic chains.

At the moment the most appropriate suggestion might be that the major role of the phospholipids (in particular or even exclusively phosphatidylserine) involved in ATPase activity is the maintenance of the correct spatial orientation of the enzyme essential for its activity.

A definite conclusion regarding correct function of the compounds activating the transport ATPase cannot be made yet.

REFERENCES

1. Morton, R.K., in "Methods in Enzymology", Vol. 1, p. 25, Academic Press, New York (1955).
2. Maddy, A.H., Biochim. Biophys. Acta 88:448 (1964).
3. Maddy, A.H., Biochim. Biophys. Acta 117:193 (1966).
4. Zwaal, R.F.A. and van Deenen, L.L.M., Biochim. Biophys. Acta 163:44 (1968).
5. Zwaal, R.F.A., Studies on Membrane Proteins of Erythrocytes and their Interaction with Lipids, Thesis, University of Utrecht (1970).
6. Rega, A.F., Weed, R.J. and Rothstein, A., Federation Proc. 25:290 (1966).
7. Rega, A.F., Weed, R.J., Reed, C.F., Berg, G.G. and Rothstein, A., Biochim. Biophys. Acta 147:297 (1967).
8. Dodge, C.T., Mitchell, C.D. and Hanahan, D.J., Arch. Biochem. Biophys. 100:119 (1963).
9. Zwaal, R.F.A. and van Deenen, L.L.M., Biochim. Biophys. Acta 150:323 (1968).
10. Kamat, V.B., Chapman, D., Zwaal, R.F.A. and van Deenen, L.L.M., Chem. Phys. Lipids, in Press.
11. Green, D.E., Murer, E., Hultin, H.O., Richardson, S.H., Salmon, B., Brierly, G.P. and Baum, H., Arch. Biochem. Biophys. 112:635 (1965).
12. Rutter, W.J., Hurisley, J.R., Groves, W.E., Caldor, J., Rajkumar, T.V. and Woodfin, B.M., in "Methods in Enzymology", Vol. 9, p. 479, Academic Press, New York (1966).
13. Allison, W.S., in "Methods in Enzymology", Vol. 9, p. 210, Academic Press, New York (1966).

14. Van den Berg, J.W.O., On the Acylation of Lysolecithins by Erythrocyte Membranes, Thesis, University of Utrecht (1969).
15. Michel, H.O., J. Lab. Clin. Med. 34:1564 (1949).
16. Mitchell, C.D. and Hanahan, D.J., Biochemistry 5:51 (1966).
17. Burger, S., Fujii, T. and Hanahan, D.J., Biochemistry 7:3682 (1968).
18. Post, R.L., Merritt, C.R., Kinsolving, C.R. and Albright, C.D., J. Biol. Chem 235:1796 (1960).
19. Roelofsen, B., Some Studies on the Extractability of Lipids and the ATPase Activity of the Erythrocyte Membrane, Thesis, University of Utrecht (1968).
20. Roelofsen, B., de Gier, J. and van Deenen, L.L.M., J. Cell. and Comp. Physiol. 63:233 (1964).
21. Parpart, A.K., J. Cell. and Comp. Physiol. 19:248 (1942).
22. Solubility of Inorganic and Organic Compounds, (H. Stephen, ed.), Vol. 1, part I, p. 405, 418, Pergamon Press, Oxford (1963).
23. Somogyi, J., Naturwissenschaften 51:134 (1964).
24. Marchesi, V.T. and Palade, G.E., J. Cell Biol. 35:385 (1967).
25. Jørgensen, P.L. and Skou, J.C., Biochem. Biophys. Res. Comm. 37:39 (1969).
26. Israel, Y., in "The Molecular Basis of Membrane Function" (D.C. Tosteson, ed.), p. 529, Prentice Hall, New Jersey (1969).
27. Parpart, A.K. and Ballentine, R., in "Modern Trends in Physiology and Biochemistry" (E.S.G. Barron, ed.), Academic Press, New York (1952).
28. Lester, R.L. and Fleischer, S., Biochim. Biophys. Acta 47:358 (1961).
29. Roelofsen, B., de Gier, J. and van Deenen, L.L.M., Proc. Koninkl. Ned. Akad. Wetensch. Series B, 68:249 (1965).
30. Roelofsen, B., Baadenhuysen, H. and van Deenen, L.L.M., Nature 212:1379 (1966).
31. Schatzmann, H.J., Nature 196:677 (1962).
32. Tatibana, M., J. Biochem. 53:260 (1963).
33. Ohnishi, T., Nakamura, Y. and Kawamura, H., J. Phys. Soc. Japan 19:420 (1964).
34. Tanaka, R. and Abood, L.G., Arch. Biochem. Biophys. 108:47 (1964).

35. Ohnishi, T. and Kawamura, H., J. Biochem. 56:377 (1964).
36. Tanaka, R. and Strickland, K.P., Arch. Biochem. Biophys. 111:583 (1965).
37. Fenster, L.J. and Copenhaver Jr., J.H., Biochim. Biophys. Acta 137:406 (1967).
38. Tanaka, R., J. Neurochem. 16:1301 (1969).
39. Hegyvary, G. and Post, R.L., in "The Molecular Basis of Membrane Function" (D.C. Tosteson, ed.), p. 519, Prentice Hall, New Jersey (1969).
40. Wheeler, K.P. and Whittam, R., Nature 225:449 (1970).
41. De Haas, G.H., Postema, N.M., Nieuwenhuizen, W. and van Deenen, L.L.M., Biochim. Biophys. Acta 159:103 (1968).
42. Sanders, H., Biochim. Biophys. Acta 144:485 (1967).
43. Tanaka, R. and Sakamoto, T., Biochim. Biophys. Acta 193:384 (1969).

SIMILARITIES AND DISSIMILARITIES BETWEEN OUTER MITOCHONDRIAL MEMBRANE AND ENDOPLASMIC RETICULUM

Gian Luigi Sottocasa

Istituto di Chimica Biologica - Università di Trieste

Trieste, Italy

At the onset of this discussion it is important to define what I mean by endoplasmic reticulum and outer mitochondrial membrane. As most of biochemists I use the term endoplasmic reticulum as referred to the microsomal fraction obtained by conventional differential centrifugation at 105,000 x g for one hour, after sedimentation of mitochondria. This fraction is heterogeneous and can be subfractionated by different means (1,2). The distribution of certain enzymes, however, can be regarded as sufficiently homogeneous throughout the different endoplasmic reticulum vescicles to be considered safe the use of such enzymes as markers of microsomes. These enzymes include glucose-6-phosphatase, NADPH cytochrome c reductase and a number of NADPH-linked functions of microsomes as well as cytochrome P_{450} (3). By outer mitochondrial membrane, on the other hand, I mean a fraction derived from isolated mitochondria, subjected to treatments capable to detach the outer from the inner membrane and subsequently fractionated either by continuous or discontinuous density-gradient as well as by differential centrifugation (3-12).

The study of the properties of microsomes and outer mitochondrial membrane has been undertaken in different laboratories and has revealed that enzymes such as NADH cytochrome b_5 reductase, cytochrome b_5 and some enzymes involved in phospholipid synthesis are common to the two structures. Other enzymes, on the contrary, are exclusive complement either of microsomes or outer mitochondrial membrane (see for review ref. 13). Cytochrome P_{450} is an exclusive microsomal component in liver; in other tissues, such as adrenal cortex, the pigment has been reported also in mitochondria (14,15). We found therefore of interest to investigate the intramitochondrial localization of the cytochrome in adrenal cortex. Pre-requisit for such a study was to find suitable markers for microsomes, on one hand, and for the outer mitochondrial membrane, on the other. A subfractionation of adrenal cortex mitochondria had never been undertaken before and such parameters were not known. Glucose-6-phosphatase is an excellent marker of microsomes in liver; we decided to investigate if this enzyme was present also in adrenal cortex and to study its distribution. Table I shows the distribution of glucose-6-phosphatase, NADH-cytochrome c reductase and succino-oxidase between mitochondria and microsomes. It appears that glucose-6-phosphatase shows a typical microsomal distribution. On the basis of this parameter it is possible to calculate that the microsomal contamination of mitochondria is almost negligible. Rotenone-insensitive NADH-cytochrome c reductase, an enzyme present in liver mostly in microsomes but also in the outer mitochondrial membrane, here is equally distributed between mitochondrial and microsomal fractions. On the basis of succino-oxidase activity, the contamination of the microsomal fraction by inner mitochondrial membrane may be estimated to be of the order of 1%. From the biochemical point of view mitochondria were rather well preserved, as indicated by P/O and respiratory control ratios. In addition, (not reported in the table) they were capable of oxidizing a number of NAD-linked substrates such as pyruvate, malate, α-keto-glutarate and isocitrate. Glutamate and β-hydroxybutyrate were very poor substrates for these mitochondria. We considered this mitochondrial preparation suitable to attempt the separation of the two membranes. Direct application of the technique used for liver mitochondria (3) gave, however, unsatisfactory results. The reason is essentially that mitochondria from adrenal cortex differ from liver mitochondria in that they show a lower equilibrium density and are less susceptible to sonic oscillation. The best conditions for the rupture and detachment of the outer membrane are listed in the following scheme:

<u>Swelling</u>: 10 mM Tris-Pi (pH 7.5), 5 minutes at 0°C
<u>Shrinkage</u>: 0.9 M sucrose + 0.5 mM ATP, 5 minutes at 0°C
<u>Fragmentation</u>: sonic oscillation (Branson) 3 amp, 40", 0°C
<u>Centrifugation</u>: linear density-gradient between 0.7 M and 1.54 M sucrose containing 0.1 mM EDTA (d= 1.100 - 1.205) 4 hrs, Spinco SW-25 rotor, 24,000 RPM.

TABLE I

Some Enzyme Activities of Mitochondria and Microsomes Isolated from Bovine Adrenal Cortex

	G-6-Pase µmoles Pi/20 min /mg prot.	Rot.-insensitive NADH cyt.c red. µmoles cyt.c red. /min/mg prot.	Succinate oxidation		
			μAO_2/min/mg prot.	P/O	R.C.
Mitochondria	0.019	0.473	0.247	1.88	5.0
Microsomes	0.330	0.478	0.003	--	--
Mitochondria / Microsomes	0.057	0.990	82	--	--

As shown in Fig. 1, the treatment results in the separation of two major protein bands: a light one, containing the bulk of rotenone-insensitive NADH cytochrome c reductase virtually devoid of succinate cytochrome c reductase, and a heavy one collecting the bulk of the latter enzyme, practically devoid of the former. The obvious conclusion one may draw from these data is that, similarly to liver-mitochondria, the outer membrane in this tissue contains an active NADH cytochrome c reductase insensitive to rotenone. As expected, the spectral analysis of the fraction has revealed the presence of cytochrome b_5, which is specifically reduced by NADH. The appearance of the two fractions at the electron microscope was indistinguishable from similar fractions obtained from liver-mitochondria. The only appreciable difference was in the size of the vescicles of the outer membrane which in this case were much smaller. This finding is not surprising in view of the more drastic sonication conditions applied.

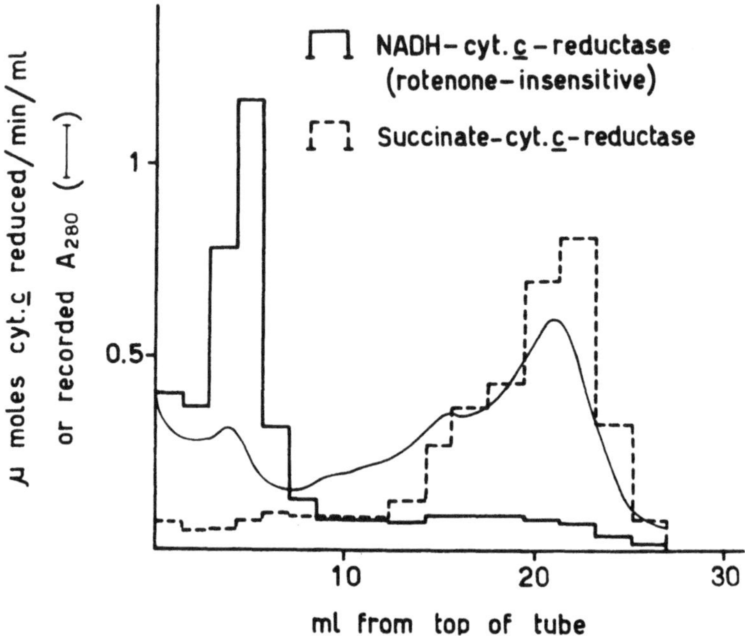

Fig. 1. Distribution of rotenone-insensitive NADH cytochrome c reductase, succinate cytochrome c reductase and total protein in subfractions derived from adrenal cortex mitochondria in a linear density-gradient.

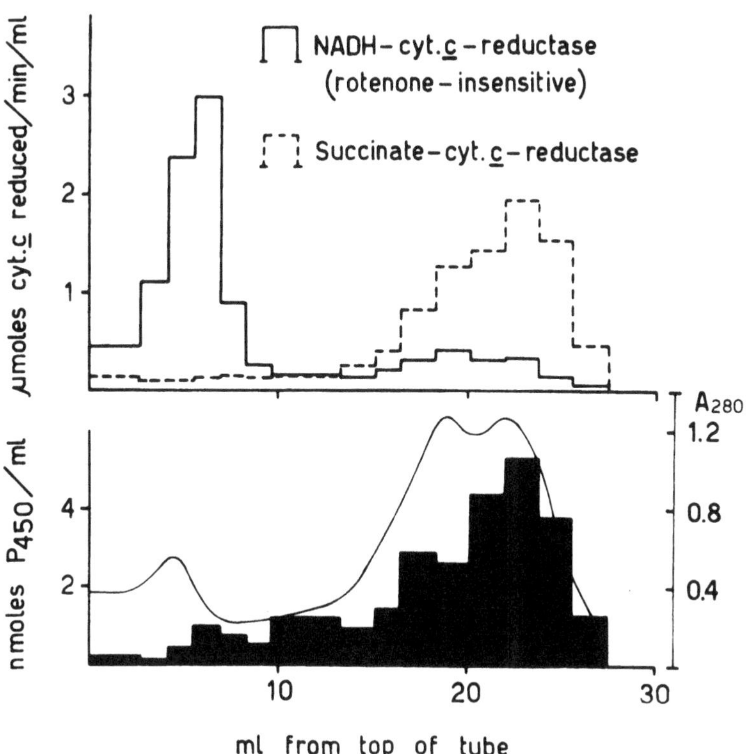

Fig. 2. Distribution of rotenone-insensitive NADH cytochrome c reductase, succinate cytochrome c reductase, cytochrome P_{450} and total protein in subfractions derived from adrenal cortex mitochondria in a linear density-gradient.

The distribution of cytochrome P_{450} and other enzymic parameters are reported in Fig. 2. Clearly the distribution of the cytochrome follows that of succinate cytochrome c reductase, indicating that the pigment is bound to the inner rather than to the outer mitochondrial membrane. Some of these results have already been reported (16,17) and similar findings have been reached independently in other laboratories (18). The obvious conclusion is that cytochrome P_{450} is absent in outer mitochondrial membrane even in a tissue where the pigment has also a mitochondrial localization.

Similarities and dissimilarities between microsomes and outer mitochondrial membrane have been found also at the chemical level. It has been found that the total phospholipids and protein as well as individual phospholipids and cholesterol are differently distributed in microsomes and outer mitochondrial membrane (7,9,19,20). An electrophoretic investigation of solubilized proteins from smooth endoplasmic reticulum and outer mitochondrial membrane has revealed that, out of fifteen and twelve protein species respectively, three components show the same electrical mobility (21). All these findings point to substantial chemical differences between the two membrane structures.

Among the physical properties, permeability is particularly interesting. I shall present here some results which led us to the conclusion that isolated microsomes are impermeable to pyridin-nucleotides, whereas isolated outer mitochondrial membrane vescicles are not. These results have been reached during the course of an investigation aimed to elucidate the phenomenon already described (22-24) that cations produce a marked stimulation of NADH oxidation by rat-liver microsomes.

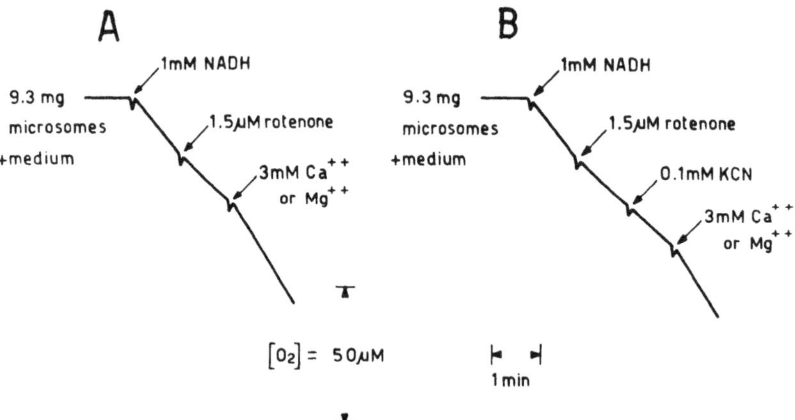

Fig. 3. Effect of calcium and magnesium on the aerobic oxidation of NADH by rat-liver microsomes.

Experimental conditions: Clark-electrode traces at 30°. Microsomes sedimented from a 10% rat-liver homogenate in 0.25 M sucrose between 15,000 x g for 15 minutes and 105,000 x g for 1 hour. The medium consisted of 0.45 M sucrose; 0.002 M tris-HCl buffer at pH 7.8. Additions as indicated in figure. Rotenone was dissolved in 3 μl ethanol; Ca^{++} and Mg^{++} were added as 1 M solutions of $CaCl_2$ and $MgSO_4$. Final volume: 0.9 ml.

As shown in fig. 3, microsomal preparations can catalyze an aerobic oxidation of NADH which is slightly inhibited by rotenone and stimulated by addition of Ca^{++} or Mg^{++}, in accordance with the data already published (22). The slight inhibition by rotenone indicates the presence of a very small contamination by inner mitochondrial membrane fragments. The extent of such a contamination was calculated to be less than 5% of the total protein, on the basis of rotenone titer (25). As shown in trace B of the figure, in the presence of rotenone, cyanide neither decreases nor prevents the effect of divalent cations. The failure of rotenone and cyanide to prevent the cation effect allows us to exclude the participation of mitochondrial respiratory chain to the phenomenon. Such a participation would also imply a stoichiometry of one mole of NADH per atom of oxygen taken up. On the contrary, the stoichiometry of NADH oxidation was found to be one mole of reduced pyridin-nucleotide per mole of oxygen taken up. The phenomenon can be explained by an effect of divalent metal ions on the apparent Km of NADH oxidation, as shown in fig. 4. A Km of about 2.6×10^{-3} M is found for the system in the absence of added metal ions. Addition of 0.3 mM Ca^{++} or Mg^{++} decreases the apparent Km to 1.3×10^{-3} M. At higher ion concentrations the decrease of Km is even more evident (0.65×10^{-3} M). Concentrations of Ca^{++} and Mg^{++} higher than 3 mM fail to further affect the Km.

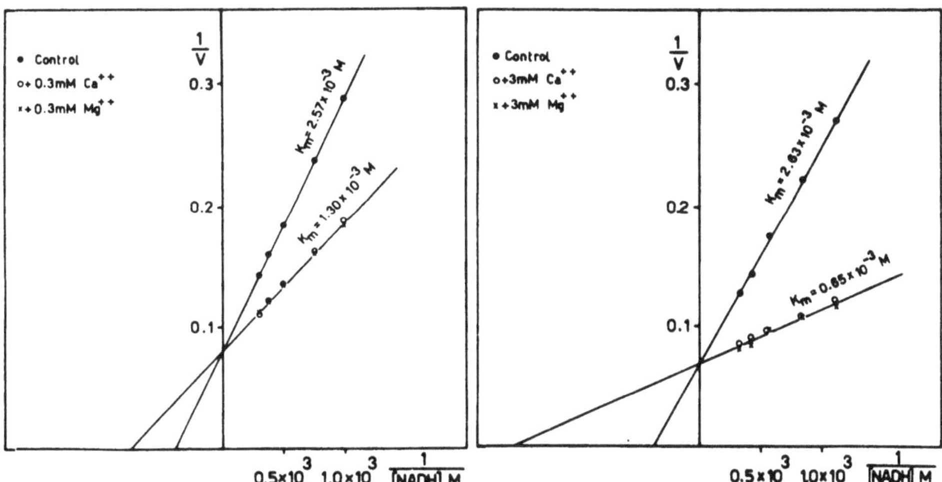

Fig. 4. Double-reciprocal plot of microsomal NADH oxidase versus NADH concentration in the presence and absence of calcium and magnesium. Experimental conditions as in fig. 3. Velocity in arbitrary units.

Similarly monovalent cations such as Na^+ or K^+ decrease the apparent Km of the system for NADH. Interestingly, however, the monovalent cations are much less efficient concentration-wise than divalent cations in decreasing the apparent Km. The extent of the maximal effect has also been found to be consistently smaller in the case of monovalent than divalent cations.

When cytochrome c rather than oxygen is the final electron-acceptor, both isolated microsomes (26) and purified NADH cytochrome b_5 reductase (27) display a Km for NADH of the order of 2.7×10^{-6}M. A discrepancy of three orders of magnitude could be explained by the following assumptions: 1) all the chains are accessible to oxygen but only a few of them can react readily with cytochrome c; 2) the chains which are readily accessible to cytochrome c are also freely accessible to NADH. These two assumptions imply that when cytochrome c is the terminal electron-acceptor only those chains which are freely accessible to both cytochrome c and NADH can operate and, on the contrary, when oxygen is the terminal acceptor, the bulk of the chains can function. It follows that the Km for NADH measured with oxygen as acceptor reflects the existence of a permeability barrier to NADH.

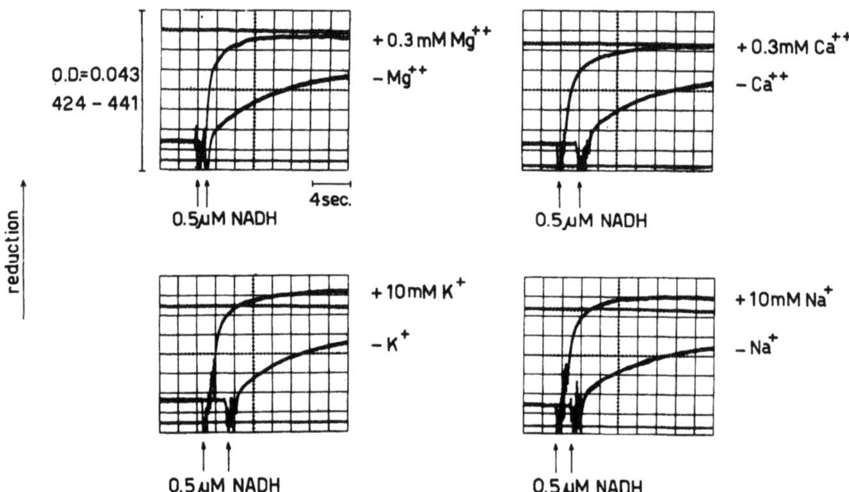

Fig. 5. Influence of cations on the rate of reduction of cytochrome b_5 in isolated rat-liver microsomes.

Experimental conditions: dual wavelength oscilloscopic traces (Phoenix spectrophotometer) at 30°. The assay mixture consisted of: 10.3 mg microsomal protein; 0.45 M sucrose; 0.003 M tris-HCl buffer at pH 7.8; 1.5 uM rotenone. Final volume 3.0 ml. Additions as indicated in figure.

The effect induced by the presence of mono- or di-valent metal ions on the Km suggests that, rather than affecting the enzyme system, metal ions would interfere with the permeability barrier, somehow facilitating the access of NADH to the active site of the enzyme. If this were the case, it is expected that addition of cations would affect both the rate of reduction and the aerobic steady-state level of reduced cytochrome $\underline{b}_5$ in microsomes. The experiments presented in fig. 5 clearly show that the rate of reduction of cytochrome $\underline{b}_5$ by NADH is strongly enhanced by addition of 0.3 mM Ca^{++}or Mg^{++}as well as by 10 mM Na^{+}or K^{+}.

The influence of the ions on the aerobic steady-state level of cytochrome $\underline{b}_5$ of microsomes is presented in fig. 6. After a certain

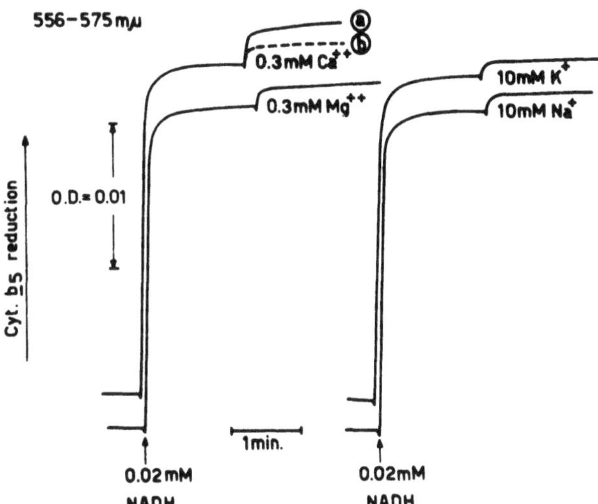

Fig. 6. Influence of cations on the aerobic steady-state level of cytochrome $\underline{b}_5$ in isolated rat-liver microsomes.

Experimental conditions as in fig. 5. Protein concentration 7.41 mg in 3.0 ml final volume.

steady-state level has been reached following the addition of 20 μM NADH, metal ions are added and a marked increase in the steady -state level ensues. In the case of Ca^{++}(trace a) the raise in the steady -state level appears greater than in the other cases, due to an increase in light scattering. A correction has therefore been made (trace b) for this artifact, based on the optical change found after the addition of the same amount of Ca^{++} in the absence of NADH. The experiments reported in figures 5 and 6 support the view that metal ions favour the access of NADH to the oxidizing system. It is worth while mentioning in this respect that recently Amons et al (28) have suggested that KCl may favour the entry of ATP into rat-liver mitochondria. Similarly it may be suggested that metal ions favour the entry of pyridin-nucleotides into microsomal vescicles.

TABLE II

Solubilisation of NADH Oxidase from Rat-liver Microsomes

The microsomal suspension in 0.45 M sucrose was diluted to 20 mg protein/ ml; 40 ul/ ml triton X-100 were added and the suspension homogenized by means of a hand-moved Potter homogenizer. The centrifugation was performed for 1 hour at 105,000 x g in the No. 40 Spinco rotor. The pellet was suspended in 0.45 M sucrose. NADH oxidase was determined as described in fig. 3.

Fraction	Total protein (mg)	NADH oxidase (uAO_2/min)		Km (M) for NADH	
		mg $prot.^{-1}$	total	+3mM Ca^{++}	$-Ca^{++}$
Microsomes	679.5	0.0040	2.70	$0.6x10^{-3}$	$2.6x10^{-3}$
Microsomes + 4% triton X-100	679.5	0.0037	2.53	--	--
Supernatant at 105,000 x g	226.0	0.0084	1.90	$2.1x10^{-6}$	$2.1x10^{-6}$
Pellet	498.0	0.0012	0.61	$0.4x10^{-3}$	$0.8x10^{-3}$
% Recovery	106		99		

To further test the hypothesis of a permeability barrier to NADH in microsomes we have attempted the solubilisation of the enzyme sys-stem by means of triton X-100. The results of this experiment are shown in Table II. Upon treatment with the detergent, followed by high speed centrifugation, approximately 75% of the total activity is recovered in the supernatant with about 2.5-fold increase in specific activity. Washing the pellet with the same medium (not shown in the Table) increases the amount of enzyme solubilized to 85%. The enzyme activity is recovered up to 99% after separation. It may be noticed that the Km for NADH found for the soluble fraction has been decreased by the procedure three orders of magnitude in contrast with the enzyme still bound to the sedimentable fraction which displays a Km of the same order of magnitude as the untreated microsomes. As expected, addition of divalent cations does not affect the Km in the case of the soluble fraction and decreases the Km of the sedimentable enzyme. These results support the view that indeed a permeability barrier exists for NADH in rat-liver microsomes. In addition, similar results have been reached by solubilisation of a completely independent enzyme such as NAD glycohydrolase.

All the results so far presented support the view that there exists a permeability barrier to pyridin-nucleotides in microsomes and, presumably, also in the endoplasmic reticulum from which they derive.

If the Km for NADH of NADH oxidase and the influence on this parameter by metal ions can be used as criterion to decide whether the endoplasmic reticulum membrane is permeable to exogenous NADH, then the determination of such a parameter with the outer mitochondrial membrane could be useful to decide if this membrane is permeable to pyridin-nucleotides intrinsically or its permeability to the compounds, in isolated mitochondria, reflects damage occurring during isolation. It is expected, in fact, that the outer mitochondrial membrane separated by swelling, shrinking and sonication, being constituted of small vescicles would retain its permeability properties similarly to microsomes which are also round vescicles resealed after disruption of the endoplasmic reticulum.

Polarographic traces in fig. 7 show some properties of the aerobic oxidation of NADH catalyzed by preparations of outer mitochondrial membrane. In the presence of rotenone (trace a) the rate of oxygen uptake is greatly stimulated by addition of divalent cations such as Ca^{++} or Mg^{++}. The oxygen uptake, however, is strongly inhibited by addition of cyanide. Trace b of the figure shows that, in contrast to microsomes, oxygen uptake of outer mitochondrial membrane preparations is not stimulated by divalent cations in the presence of potassium cyanide. The obvious explanation of these findings is that the preparation is contaminated by a catalytic amount of soluble cytochrome

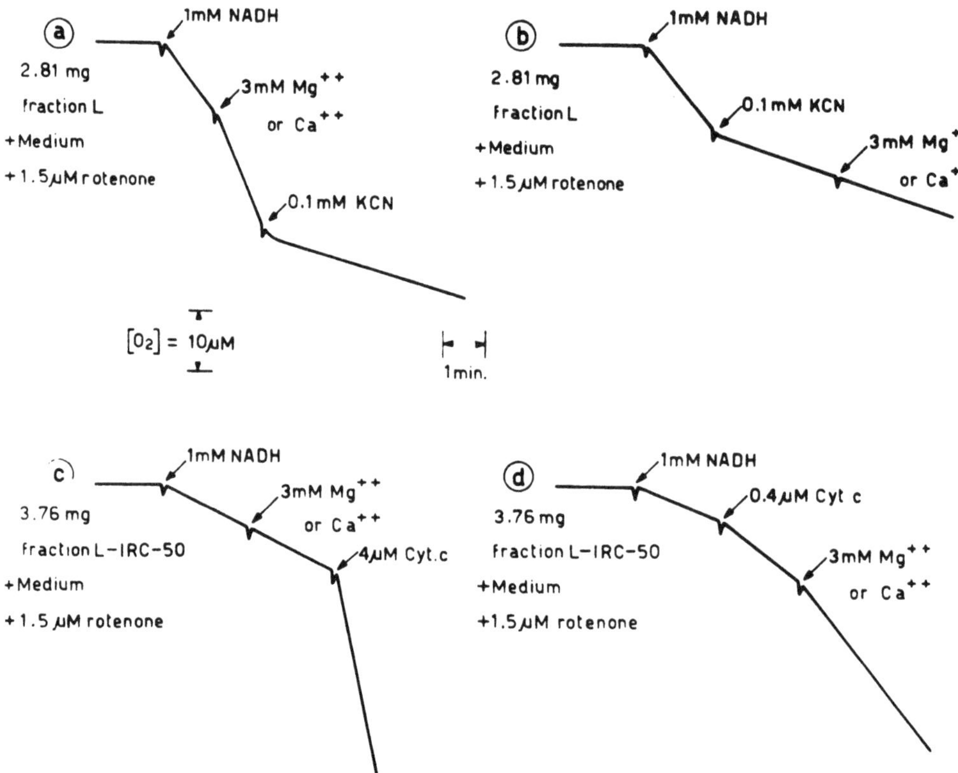

Fig. 7. Influence of divalent cation on the rate of aerobic oxidation of NADH by outer mitochondrial membrane preparations from liver. Experimental conditions: the outer mitochondrial membrane (fraction L in the figure) was prepared as described by Sottocasa et al (3) according to the modification reported by Sottocasa and Sandri (29). Mg^{++} was omitted from the contracting medium. Fraction L-IRC-50 was obtained by ion-exchange chromatography of fraction L on a short column (0.5 x 10 cm) of Amberlite IRC-50, sodium form, pre-equilibrated with 0.45 M sucrose, containing 0.010 M phosphate buffer at pH 7.2. Assay medium as in fig. 3.

c which mediates the electron flow from cytochrome b_5 to cytochrome-oxidase of inner mitochondrial membrane fragments, also present as contaminants in the preparation. If this is the case, removal of cytochrome c should result in abolishing the effect of cations on the rate of oxygen uptake. Experiments recorded in traces c and d show that, when the outer mitochondrial membrane fraction is subjected to ion-exchange chromatography on Amberlite IRC-50 prior to use, no effect of Ca^{++}or Mg^{++} is,in fact detectable. The effect of divalent metal ions can be restored by adding a trace amount of cytochrome c. The results of this experiment are in contrast with those found with microsomes and would indicate that either a barrier to pyridin-nucleotides does not exist in the outer mitochondrial membrane or the membrane is insensitive to addition of cations. To decide between these two alternatives we have determined the Km for NADH of the system. Accurate measurements of this parameter gave a result of 1.97 ± 0.02 x 10^{-6} M. This value clearly indicates that no barrier can exist between the pyridin-nucleotide and the active site of the enzyme in the outer mitochondrial membrane. This conclusion is supported also by the results of the experiments presented in fig. 8. Upon addition of Ca^{++} the aerobic steady-state level of reduced cytochrome b_5 does not increase appreciably in outer mitochondrial membrane preparations. The change in absorption upon addition of the ion, shown in the left hand trace of the figure, is obviously due to light-scattering change as indicated by the experiment presented in the right hand trace.

All these findings are in good agreement with the almost generally accepted idea that the outer mitochondrial membrane is freely permeable to nucleotides and, in addition, suggest that this property is intrinsic in the membrane and not due to the presence of discontinuities produced during the isolation of mitochondria. If this were, in fact, the case it is expected that upon sonication the re-arrangement of the fragments of the outer mitochondrial membrane would result in a preparation of small vescicles essentially impermeable to

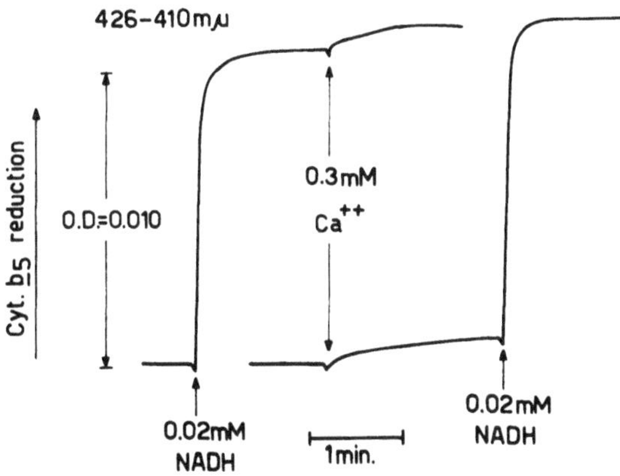

Fig. 8. Influence of calcium on the aerobic steady-state level of cytochrome b_5 in isolated outer mitochondrial membrane from rat-liver. Experimental conditions as in fig. 5, in the presence of 3.8 mg outer membrane protein and 0.1 mM potassium cyanide.

nucleotides.

An alternative explanation might be based on three assumptions: a) the outer mitochondrial membrane is damaged in isolated mitochondria; b) the outer mitochondrial membrane is asimmetric with NADH cytochrome $\underline{b}_5$ system sitting on the inner surface; and c) sonication results in an inversion of the vescicles which at the end are therefore inside-out. The latter explanation seems far more complicated though not impossible.

The knowledge so far reached concerning the enzymic, chemical and physical properties of isolated microsomes and outer mitochondrial membrane seems to indicate that these two intracellular structures, though not completely dissimilar, differ substantially from one another. Studies of the turnover of individual components of the two membranes have also revealed different turnover rates (30-36). Particularly the turnover rate of cytochrome $\underline{b}_5$ isolated from the two membranes has been found to be faster in microsomes (2.5 days) than in the outer mitochondrial membrane (4.4 days) (37). The latter findings together with the above mentioned results do not role out the possibility of a derivation of the outer membrane from the endoplasmic reticulum. The two membranes, however, have their own individuality and behave as distinct entities.

REFERENCES

1. Dallner, G., Acta Pathol. et Microbiol. Scand. Suppl. 166:1 (1963).
2. Dallman, P.R., Dallner, G., Bergstrand, A., and Ernster, L., J. Cell Biol. 41:357 (1969).
3. Sottocasa, G.L., Kuylenstierna, B., Ernster, L. and Bergstrand, A., J. Cell Biol. 32:415 (1967).
4. Sottocasa, G.L., Kuylenstierna, B., Ernster, L. and Bergstrand, A., in "Methods in Enzymology" (R.W. Estabrook and M.E. Pullman, eds.), Vol. 10, p 448, Acad. Press, New York (1967).
5. Sottocasa, G.L., Ernster, L., Kuylenstierna, B. and Bergstrand, A., in "Mitochondrial Structure and Compartmentation" (E. Quagliariello, S. Papa, E.C. Slater and J. M. Tager, eds.), p 74, Adriatica Editrice, Bari (1967).
6. Parsons, D.F., Williams, G.R. and Chance, B.,Ann. N. Y. Acad. Sci. 137:643 (1966).
7. Parsons, D.F., Williams, G.R., Thompson, W., Wilson, D. and Chance, B., in "Mitochondrial Structure and Compartmentation" (E. Quagliariello, S. Papa, E.C. Slater and J.M. Tager, eds.), p 29, Adriatica Editrice, Bari (1967).

8. Parsons, D.F., and Williams, G.R., in "Methods in Enzymology" (R.W. Estabrook and M.E. Pullman, eds.), Vol. 10, p 443, Acad. Press, New York (1967).
9. Lèvy, M. and Sauner, M-T., Compt. Rend. Soc. Biol., 161:277 (1967).
10. Lèvy, M., Toury, R. and Andrè, J., C. R. Acad. Sci. Paris, Sér. D, 263:1766 (1966).
11. Schnaitman, C.A., Erwin, V.G. and Greenawalt, J.W., J. Cell Biol., 32:719 (1967).
12. Schnaitman, C.A. and Greenawalt, J.W., J. Cell Biol., 38:158 (1968).
13. Ernster, L. and Kuylenstierna, B., in "Membranes of Mitochondria and Chloroplasts" (E. Racker, ed.), p 172, Van Nostrand Reinhold Co., New York (1970).
14. Harding, B.W., Wong, S.H. and Nelson, D.H., Biochim. Biophys. Acta, 92:415 (1964).
15. Omura, T., Sato, R., Cooper, D.Y., Rosenthal, O. and Estabrook, R.W., Fed. Proc., 24:1181 (1965).
16. Sottocasa, G.L. and Sandri, G., Abstr. Int. Symp. on Microsomes and Drug Oxidation, Tübingen (1969).
17. Sottocasa, G.L. and Sandri, G., Biochem. J., 116: 16p (1970).
18. Satre, M., Vignais, P.V. and Idelman, S., FEBS Letters, 5:135 (1969).
19. Stoffel, W. and Schiefer, H-G., Hoppe Seyler's Z. Physiol. Chem., 349:1017 (1968).
20. Parsons, D.F. and Yano, Y., Biochim. Biophys. Acta, 135:362 (1967).
21. Schnaitman, C.A., Proc. Nat. Acad. Sci., 63:412 (1969).
22. Sottocasa, G.L., in "Membrane Models and the Formation of Biological Membranes" (L. Bolis and B.A. Pethica, eds.), p 229, North Holland Publ. Co., Amsterdam (1968).
23. Sottocasa, G.L. and Sandri, G., Abstr. 5th FEBS Meeting, Prague, p 131 (1968).
24. Sottocasa, G.L. and Sandri, G., in "Biochemistry of Intracellular Structures: Mitochondria and Endoplasmic Reticulum" (L. Wojtczak, W. Drabikowski and H. Strzelecka-Golaszewska, eds.), p 7, P. W. N. Warszawa (1969).
25. Ernster, L., Dallner, G. and Azzone, G.F., J. Biol. Chem., 238: 1124 (1963).
26. Kuylenstierna, B. and Sottocasa, G.L., umpublished data.
27. Strittmatter, P. and Velick, S.F., J. Biol. Chem., 228:785 (1957).
28. Amons, R., Van den Bergh, S.G. and Slater, E.C., Biochim. Biophys. Acta, 162:452 (1968).

29. Sottocasa, G.L. and Sandri, G., It. J. Biochem., 17:17 (1968).
30. Omura, T., Siekevitz, P. and Palade, G.E., J. Biol. Chem., 242: 2389 (1967).
31. Brunner, G. and Neupert, W., FEBS Letters, 1:153 (1968).
32. Bygrave, F.L. and Bücher, Th., European J. Biochem., 6:256 (1968).
33. Bygrave, F.L. and Bücher, Th., FEBS Letters, 1:42 (1968).
34. Bygrave, F.L., J. Biol. Chem., 244:4768 (1969).
35. Beattie, D.S., Biochem. Biophys. Res. Commun., 35:721 (1969).
36. de Bernard, B., Getz, G.S. and Rabinowitz, M., Biochim. Biophys. Acta, 193:58 (1969).
37. Druyan, R., de Bernard, B. and Rabinowitz, M., J. Biol. Chem., 244:5874 (1969).

HORMONAL INTERACTIONS AND REGULATION OF ADENYLCYCLASE ACTIVITY IN ISOLATED LIVER PLASMA MEMBRANE

Vittorio TOMASI, Agostino TREVISANI and
Ottavio BARNABEI

Institute of General Physiology, University of Ferrara, Ferrara, Italy

INTRODUCTION

The idea that hormones may act primarily at the surface of the cell to influence cell function by way of regulating the permeability to metabolites or ions (or their transport), is probably the oldest theory of hormone action. It was once very popular, then it was obscured by the more attractive theories that hormone may act by modifying the activity of particular enzyme systems (hormone-enzyme thesis) or their synthesis (hormone-gene or hormone-protein synthesis thesis). At present, however, the view that cell membrane may represent a site of primary action for many hormones is growing again popular. Several new concepts have favoured this re-discovery. The cell membrane is now regarded as an assembly of lipoprotein subunits, lipid and protein being linked mainly by hydrophobic bonds (1-3). Emphasis in this new picture lays greater stress on the action of membrane protein and enzymes rather than the lipids to explain both the transport mechanisms and the hormonal interactions. Moreover it has become evident that transport processes may depend on enzyme activities and conversely that enzyme activities may depend on the association of enzymes with membranes.

However, the most impressive evidence about the role of cell membrane in hormonal response is provided by the work of Sutherland and his collaborators (4), who have opened new biochemical perspectives with their discovery of 3',5'-cyclic AMP and their suggestion that this compound may act as a second messenger, the function of which is to propagate throughout the cell the message brought by the first messenger, the hormone. The ultimate response depends on the enzymatic profile of the type of cell involved.

At present numerous hormones, including epinephrine, glucagon, ADH, TSH, ACTH, parathormone and several others are believed to exert, at least some of their effects, by the "second messenger" mechanism, i.e. by causing an alteration in the intracellular level of cyclic AMP. This imply that cyclic AMP may play a variety of roles. To list only some, it is considered a mediator in hormonal effects on muscle and liver phosphorylase and glycogenosynthetase, on gluconeogenesis in liver, on steroidogenesis in the adrenal cortex and the ovary, on the permeability of the toad bladder and collecting tubules in the kidney (see 4 and 5).

The intracellular levels of cyclic AMP are controlled by two key enzymes: adenylcyclase and phosphodiesterase. The latter is located partly in the particulate material and partly in the soluble fraction, whereas adenylcyclase is assumed to be a constituent of cell membrane (4,6). Therefore presumably, there is no need for an hormone to penetrate beyond the cell membrane for modifying the level of cyclic AMP. It can be assumed that it may interact with a membrane sited receptor, which may be either a part of the adenylcyclase molecule, as suggested by Sutherland *et al.* (4) and Robison *et al.* (6), or a component of a macromolecular assembly coupled spatially with the enzyme, as suggested by Hechter (7). Of course, it must be assumed that the highly specific receptors for the diverse hormones acting through the cyclic AMP, are either a set of adenylcyclase molecules (or moieties of them), differing in structure and specificity, or alternatively a set of other receptor molecules.

Among the hormones which are known to stimulate adenylcyclase in various tissues, the catecholamines figure prominently (4-6). Working in the department of biochemistry of Rochester, one of us (V.T.) was able to show that isolated plasma membranes of rat liver actually contains adenylcyclase activity, which was stimulated in vitro by epinephrine and glucagon (figure 1) (8).

According to the suggestion of Sutherland (4,6) this effect may represent the "primum movens" of a series of events which, through the increased formation of cyclic AMP, may led to activation of phosphorylase, glycogenolysis and hyperglicemia. As in this laboratory it has been found that the infusion of acetylcholine or carbamylcholine in the isolated, perfused rat liver produces opposite actions, namely increased deposition of glycogen and decreased output of glucose (figure 2) (9), it seemed of interest to assay the effects of acetylcholine on liver plasma membranes. The results to be described here suggest that acetylcholine reduces in vitro the activity of adenylcyclase and counteracts the stimulating effect of epinephrine.

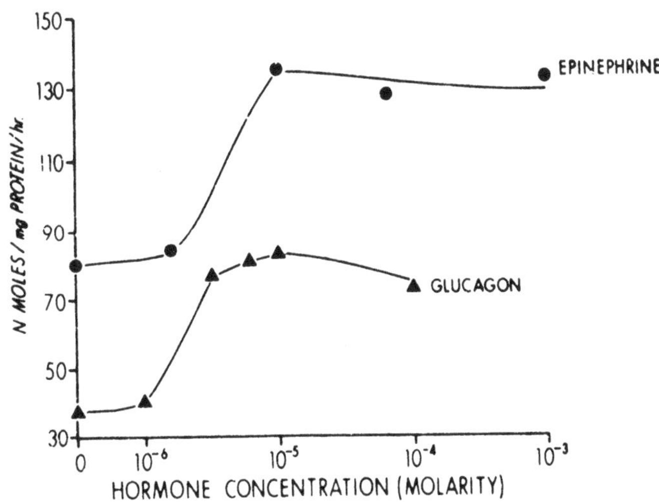

Fig. 1. The effect of epinephrine and glucagon on adenylcyclase activity of isolated plasma membranes from rat liver (from 8).

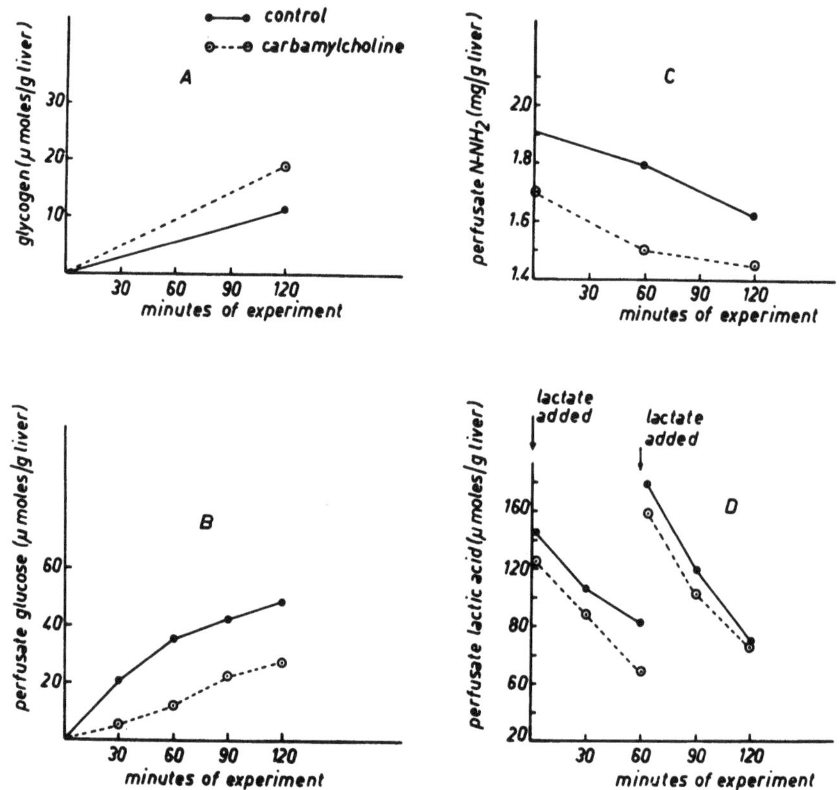

Fig. 2. Effect of carbamylcholine on deposition of glycogen and output of glucose, in the isolated perfused rat liver (from 9).

0.6 mg carbamylcholine was added at hourly intervals.
PANEL A : glycogen formation;
PANEL B : glucose output;
PANEL C : change of amino-nitrogen in the perfusate;
PANEL D : change of lactate in the perfusate.

EXPERIMENTAL

Liver plasma membranes were separated from fed rats, by the procedure of Ray (10), which is a modification of the method of Neville (11), the most important difference consisting in the inclusion of Ca^{++} 0.5 mM in the homogenizing medium. From 10 g of pooled livers about 10-13 mg of membrane protein was obtained. The purity of the membrane preparations was checked by electron microscopy and by the assay of marker enzymes: 5'-nucleotidase and glucose-6-phosphatase.

Adenylcyclase activity was measured according to Marinetti et al. (8); the composition of the assay medium and the experimental conditions are given in figure 3. The radioactivity of the newly formed cyclic AMP was determined in a Nuclear Chicago liquid scintillation counter. The binding and the release of labeled ^3H - epinephrine has been studied as described in table 1; the binding of ^{14}C-acetylcholine has been measured as described in figure 5. 5'-nucleotidase, glucose-6-phosphatase and Mg^{++} - ATPase were assayed as described by Ray (10); phosphodiesterase was tested according to the method reported by Cheung (12). Acetylcholinesterase was assayed by two different procedures (13-14). Protein was determined by the method of Lowry et al. (15) with bovine serum albumin as a standard. Epinephrine and norepinephrine were determined by the procedure of Lund (16-17), after extracting the plasma membranes according to Potter and Axelrod (18).

RESULTS AND DISCUSSION

Characteristics of Liver Plasma Membrane.

In preliminary experiments, the characteristics of isolated plasma membranes,have been studied. The results were substantially in agreement with those reported by Ray (10). The following enzymes were found to be present in our preparations: 5'-nucleotidase; adenylcyclase; Mg^{++} - ATPase. The activity of acetylcholinesterase was below the sensitivity of the methods employed (13-14). For this reason, when adenylcyclase control by acetylcholine was studied, no inhibitor of acetylcholinesterase was added in the medium (see later). The activity of cyclic AMP-phosphodiesterase was extremely low, (16-22 nmoles Pi/mg protein/15'); that of glucose-6-phosphatase was about 0.4 µmoles/mg protein/15'.

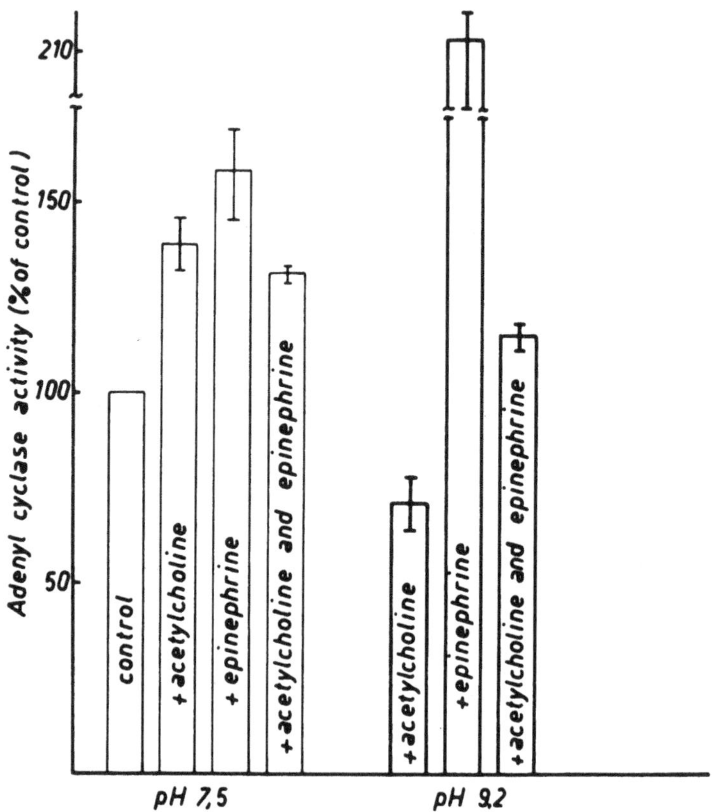

Fig. 3. Action of acetylcholine and epinephrine on adenylcyclase activity of isolated liver plasma membrane from intact rats.

The assay of adenylcyclase was as follows: liver plasma membranes were incubated for 15 min. at 37° C in a medium containing 40 mM Tris-HCl buffer; $MgCl_2$ 1.4 mM; ATP 0.8 mM; 8-^{3}H-ATP 4 μC; 50 μg serum albumin; 100-150 μg membrane protein; final volume 0.5 ml. The data are expressed in per cent of controls and are the average of 3-4 experiments. Ranges are indicated by vertical bars. The adenylcyclase activity of controls (8 experiments), as nmoles cyclic AMP formed/mg protein/hr $\pm$ S.E., was 3.81 $\pm$ 1.68 at pH 7.5 and 7.40 $\pm$ 1.39 at pH 9.2. The concentration of epinephrine (as epinephrine bitartrate) in the system was 12 μg; that of acetylcholine (as acetylcholine chloridrate), 25 μg.

The Effect of Acetylcholine and Epinephrine on Adenylcyclase Activity of Liver Plasma Membranes

The effect of acetylcholine and of epinephrine on adenylcyclase was tested at two different pH values of the medium: pH 7.5 and pH 9.2 (figure 3). Epinephrine enhanced the activity at both pH's. On the other hand, acetylcholine had an apparently paradoxical effect: at pH 9.2, it reduced the adenylcyclase activity and, added together with epinephrine, prevented its effect; however at pH 7.5 it enhanced the enzyme activity.

This result was surprising, because we normally believe that epinephrine and acetylcholine should have opposite effects; similarly nerve impulse arriving to a tissue by sympathetic fibres have the opposite effects to nerve impulses arriving by parasympathetic fibres and this is the reason by which the autonomic nerve system can be regarded as a fine homeostatic mechanism. Therefore it was reasoned that the stimulation of adenylcyclase found at pH 7.5, could not be a direct effect of acetylcholine. It is well known that the sympathetic nerves to the adrenal medulla liberate acetylcholine, which in turn evokes a secretion of catecholamines (19); more recently Burn and Rand (20) claimed that all postganglionic sympathetic nerve endings do not release norepinephrine directly, but release first acetylcholine which in turn produces a release of norepinephrine from stores which are in the region of the nerve ending by an action which is probably mediated by Ca^{++} ions.

To test the possibility that acetylcholine enhanced at pH 7.5 adenylcyclase activity by relasing epinephrine or norepinephrine, it seemed of interest to assay first the action of acetylcholine on liver plasma membranes from rats pre-treated with reserpine, which is known to deplete the catecholamines stores (21). The results, shown in figure 4, indicate that in such membranes acetylcholine produces an inhibition of the adenylcyclase activity at both pH 7.5 and pH 9.2. It seems therefore likely that when the enzyme was assayed at pH 7.5 with membranes from untreated rats, a liberation of catecholamine(s) took place, which in turn produced an increase of the activity. This point was tested by another approach. Liver plasma membranes were first incubated with labeled epinephrine. The experiment was carried out at pH 10, since it was found that the maximal binding of the hormone

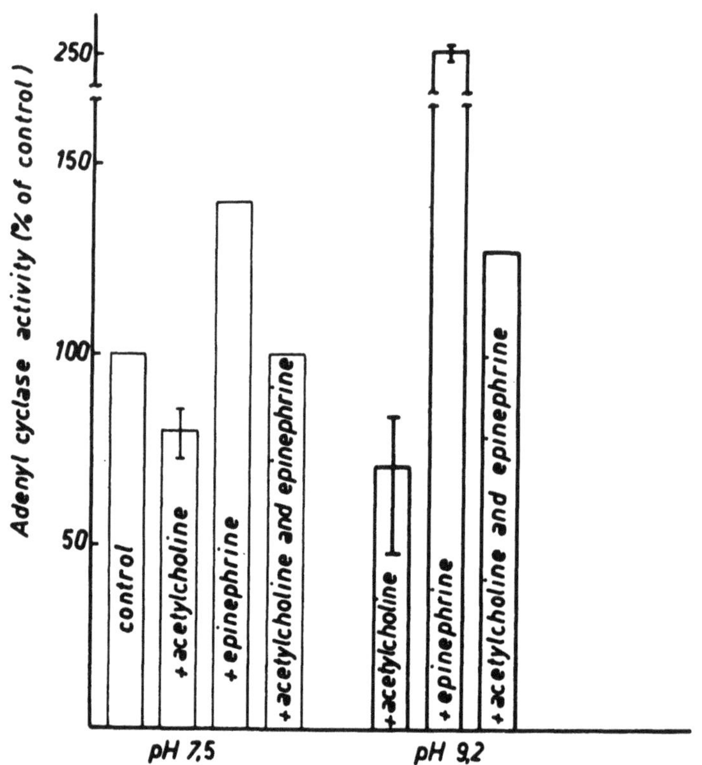

Fig. 4. Action of acetylcholine and epinephrine on adenylcyclase activity of isolated liver plasma membranes from rats pre-treated with reserpine (2.5 mg/Kg, injected intraperitoneally 48 and 24 hrs before the experiment).

For experimental conditions, see figure 3. The adenylcyclase activity of controls (4 experiments) as nmoles cyclic AMP formed/mg protein/hr ± S.E., was 5.20 ± 1.25 at pH 7.5 and 6.47 ± 1.79 at pH 9.2.

TABLE 1

Action of acetylcholine on the release of membrane bound ^{3}H-epinephrine

Experimental condition	Membrane bound epinephrine p moles/mg protein (*)	
	pH 7.5	pH 9.2
Non incubated membranes	30.4	37.6
Control	24.5	35.4
Acetylcholine 5 µg	26.6	36.7
Acetylcholine 25 µg	19.3	38.1
Bretylium(**) 50 µg	25.2	
Acetylcholine 25 µg + bretylium 50 µg	25.4	

(*) average of three experiments.

(**) as bretylium tosylate.

2-3 mg of liver plasma membrane was pre-incubated in the presence of 0.1 M glycine - NaOH buffer, pH 10, 100 µg bovine serum albumin, 2.5 µC of DL - ^{3}H-epinephrine (specific activity 6.3 C/mmole) and 5 µg of L-epinephrine bitartrate; final volume 2.2 ml. After 15 min at 37° C., the mixture was centrifuged at 3.000 x g for 10 min. the sediment was suspended in the same buffer and recentrifuged until the supernatant did not contain measurable radioactivity. The sediment was then suspended in cold 0.2 M Tris-HCl buffer, pH 7.5 or 9.2, and aliquots of the suspension (corresponding to 200 µg of proteins) were incubated 15 min at 37° C. The reaction was stopped with cold 1 N acetic acid, and the centrifuged precipitate was washed 3 times with 1 N acetic acid, and dissolved in 0.3 ml of Soluene^R. Then 15 ml of Bray's reagent were added and the radioactivity assayed.

occurs at this pH value (22). After exhaustive washings to eliminate the excess of the labeled hormone, the membranes, containing bound ^{3}H-epinephrine were re-incubated at pH 7.5, as described in table 1. In these conditions, the addition of acetylcholine resulted in an increased release of the labeled hormone, which was prevented by bretylium, a drug which at opportune concentration blocks the release of norepinephrine in nerve endings (23). If the pre-labeled membranes were incubated at pH 9.2, acetylcholine did not influence the release of ^{3}H-epinephrine.

These results induced us to test the effect of bretylium, added with and without acetylcholine,on the adenylcyclase activity of membranes from intact rats (table 2). Bretylium did not influence the enzyme activity, but enabled acetylcholine to exert its inhibitory effect at pH 7.5, as in the membranes of rats receiving reserpine (compare with figures 1 and 2).

The catecholamine concentration in isolated liver plasma membranes was determined by the procedure of Lund (16,17). Membranes from intact rats contained approximately 0.006 µg of norepinephrine and 0.018 µg of epinephrine per mg of protein; membranes from reserpinized animals contained about 1/10 of these values (the data are the average of 2 determinations, each performed on pooled membranes from 3 livers). Therefore it can be stated that acetylcholine may actually produce, within the membrane, a release of catecholamines, which in turn may enhance the adenylcyclase activity.

Finally, the action of acetylcholine on phosphodiesterase activity of membranes (as well as of the 600 x g supernatant of a liver homogenate) was tested. In both cases, the hormone did not modify appreciably the enzyme activity. This result presumably excludes the possibility that the modifications of cyclic AMP in our test conditions (see figure 3) is due to an altered rate of hydrolysis. Also the Mg^{++}- ATPase activity was not affected by acetylcholine.

TABLE 2

Action of bretylium on the modification of adenylcyclase activity in liver plasma membranes produced by acetylcholine

Treatment	Adenylcyclase activity* (nmoles/mg protein/hr)
Complete system	6.43
+ acetylcholine 25 μg	7.20
+ bretylium 50 μg	6.81
+ acetylcholine 25 μg + bretylium 50 μg	4.68

*Average of two experiments.

For experimental conditions, see the text. The enzyme was assayed at pH 7.5.

Binding of ^{14}C-Acetylcholine by Isolated Liver Plasma Membranes

The modifications of adenylcyclase activity, produced by acetylcholine, induced us to investigate a possible binding of labeled acetylcholine by the liver plasma membranes. The membranes were isolated from livers of intact rats, and were incubated in the presence of ^{14}C-acetylcholine, at different pH values of the medium, as described in figure 5. The reaction was stopped by addition of acetic acid 1 N. The precipitate was washed 3 times and assayed for radioactivity. The results indicate that the uptake of ^{14}C-acetylcholine by liver plasma membranes is pH-dependent. Two optima were observed, one at pH 5.5. and another at 9.2 (figure 5). No clear explanation can be given at present for this result. It could be thought that variation of pH modifies the uptake of acetylcholine by altering the charges of membrane consti-

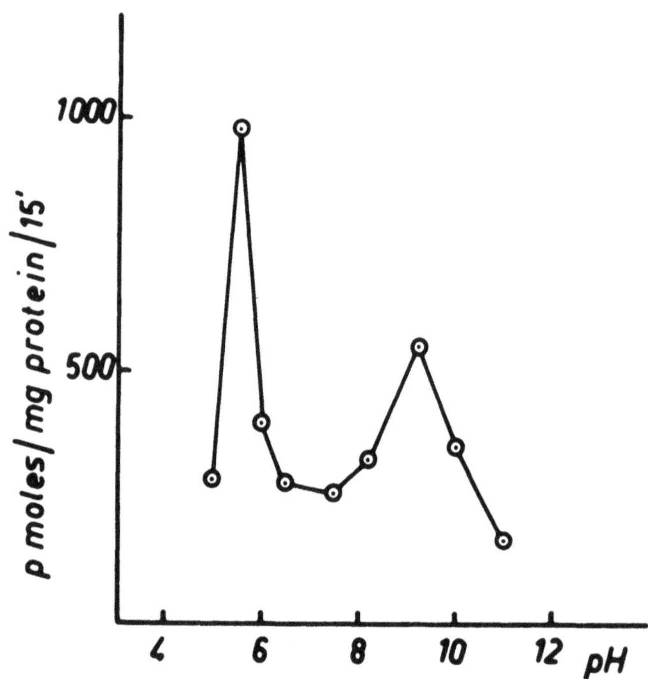

Fig. 5. Effect of pH on the binding of ^{14}C-acetylcholine by liver plasma membranes.

0.2 ml of membrane suspension (200-400 µg of membrane protein); 0.2 ml buffer; 0.1 ml of bovine serum albumin; (100 µg); 0.1 ml ^{14}C-acetylcholine (1 µC) were incubated 15' at 37° C. The reaction was stopped by addition of 2 ml of 1 N acetic acid; the precipitate was washed 3 times with 1 N acetic acid, and dissolved in 0.3 ml Soluene^{R}. Then 15 ml of Bray's reagent was added and the radioactivity assayed. The following buffers were employed: pH 5 to 6.5, 0.066 M phosphate; pH 7.5 to 9.2, 0.2 M Tris-HCl; pH 10 to 11, 0.1 M glycine-NaOH. The specific activity of the acetylcholine-1-^{14}C employed was 11.0 mC/mmole.

tuent(s) and of acetylcholine itself. However it is noteworthy that the optimum of uptake found at pH 9.2 is paralleled by the inhibition of adenylcyclase shown at the same pH value (see figures 3 and 4), which is the most constant finding observed in liver membranes from both intact and reserpine treated rats.

CONCLUSION

The concept that plasma membrane is a site of primary hormone action is at present largely accepted, after the fascinating concept of the second messenger was proposed by Sutherland and his group (4,6). The data described here suggest that the membrane may be also a site of hormonal interactions. The antagonistic effects of epinephrine and acetylcholine on glycogenesis and glucose output of liver, in fact, are paralleled by opposite actions on a key enzyme of liver plasma membrane, the adenylcyclase. It can presumably be assumed that the modification of the activity of this enzyme produces the successive, amplified responses of liver: increased glycogenesis and decreased glucose output after acetylcholine treatment; increased glycogenolysis and glucose output after epinephrine.

However, in our opinion, an important difference emerges from our data: epinephrine enhances the activity of adenylcyclase in every conditions so far tested, whereas acetylcholine may reduce, by a direct action, or enhance it, by indirect action due to release of catecholamine(s). As seen above, the content of membranes in catecholamines and the pH of medium are essential factors in the type of response produced by acetylcholine. Low concentrations of catecholamines(as in membranes from rats receiving reserpine) or a shift of pH toward higher values favour the direct action, hence the inhibition of adenylcyclase. On the other hand, lower pH's and higher content of catecholamines favour the indirect effect, i.e. the release of catecholamine(s) and the consequent activation of adenylcyclase. Probably this is a reason by which no clear effects of acetylcholine on adenylcyclase activity of crude liver preparations have been found by previous workers (24).

SUMMARY

The effect of epinephrine and acetylcholine on adenylcyclase activity of isolated plasma membrane from rat liver has been studied. Epinephrine enhanced the activity in all conditions so far tested. The effect of acetylcholine was strictly pH-dependent; when the enzyme was assayed at pH 9.2, it acted as an inhibitor, whereas it enhanced the activity at pH 7.5. On the other hand, acetylcholine at both pH's reduced the adenylcyclase activity of membranes from rats pre-treated with reserpine. It seems therefore likely that acetylcholine may inhibit the enzyme, by a direct action, or activate it, by releasing catecholamine(s). To confirm this point, isolated plasma membranes were incubated at pH 10 with ^{3}H-epinephrine; then the labeled membranes were incubated at pH 7.5. In this condition a release of bound epinephrine took place, which was enhanced by acetylcholine. Bretylium prevented this effect, and consequently elicited the inhibitory action of acetylcholine on adenylcyclase of membranes from intact rats, at pH 7.5. Finally it was found that isolated liver plasma membrane is able to bind ^{14}C-acetylcholine.

REFERENCES

1. Green, D.E. and Goldberger, F., "Molecular Insights into the Living Process", Academic Press, New York (1967).
2. Sjöstrand, F.S., in "Structural and Functional Aspects of Lipoproteins in Living Systems" (E. Tria and A. Scanu, eds.), p. 73, Academic Press, New York (1969).
3. Tria, E. and Barnabei, O., in "Structural and Functional Aspects of Lipoproteins in Living Systems" (E. Tria and A. Scanu, eds.), p. 144, Academic Press, New York (1969).
4. Sutherland, E.W., Robison, G.A. and Butcher, R.W., Circulation 38:279 (1968).
5. Munck, A., in "Recent Advances in Endocrinology", (V.H.T. James, ed.), p. 139, Churchill, London (1968).

6. Robison, G.A., Butcher, R.W. and Sutherland, E.W., Ann. N.Y. Acad. Sci. 139:703 (1967).
7. Hetcher, O., in "Mechanism of Hormone Action" (P. Karlson, ed.), p. 61, Academic Press, New York (1965).
8. Marinetti, G.V., Ray, T.K. and Tomasi, V., Biochem. Biophys. Res. Comm. 36:185 (1969).
9. Ottolenghi, C., Caniato, A. and Barnabei, O., manuscript in preparation.
10. Ray, T.K., Biochim. Biophys. Acta 196:1 (1970).
11. Neville, D.M.,Jr., J. Biophys. Biochem. Cytol. 8:413 (1960).
12. Cheung, W.J., Biochemistry 6:1079 (1967).
13. Pilz, W., in "Methods of Enzymatic Analysis" (H.V. Bergmeyer, ed.), p. 765, Academic Press, New York (1963).
14. Elmann, G.L., Courtney, D., Andres, V. and Feathrstone, R.M., Biochem. Pharmacol. 7:88 (1961).
15. Lowry, O.H., Rosebrough, N.J., Farr, A.L. and Randall, R.J., J. Biol. Chem. 193:265 (1951).
16. Lund, A., Acta Pharmacol. Toxicol. 6:137 (1950).
17. Crout, J.R., in "Standard Methods of Clinical Chemistry" (D. Seligson, ed.), Vol. 3, p. 62, Academic Press, New York (1961).
18. Potter, L.T. and Axelrod, J., J. Pharmacol. Exptl. Ther. 142:261 (1963).
19. Tepperman, J., "Metabolic and Endocrine Physiology", Year Book Medical Publisher, Chicago (1968).
20. Burn, J.H. and Rand, M.J., Ann. Rev. Pharmacol. 5:163 (1965).
21. Von Euler, U.S. and Lishajko, F., in "Pharmacology of Adrenergic and Cholinergic Transmission" (G.B. Koelle, W. Douglas and A. Carlsson, eds.), p. 245, Pergamon Press, New York (1965).
22. Tomasi, V., Koretz, S., Ray, T.K., Dunnick, J. and Marinetti, G.V., Biochim. Biophys. Acta in press.
23. Kopin, I.J., Ann. N.Y. Acad. Sci. 144:558 (1967).
24. Murad, F., Chi, Y.M., Rall, T.W. and Sutherland, E. W., J. Biol. Chem. 237:1233 (1962).

AUTHOR INDEX

(Underscored numbers indicate complete paper in this volume. Numbers followed by an asterisk refer to the pages on which the complete references are listed).

SUBJECT INDEX

www.ingramcontent.com/pod-product-compliance
Ingram Content Group UK Ltd.
Pitfield, Milton Keynes, MK11 3LW, UK
UKHW051127260726
13967UKWH00010B/2901

* 9 7 8 1 4 6 1 4 4 6 1 7 0 *